The Story of Trees

LAURENCE KING

Published in 2020
by Laurence King Publishing Ltd
361–373 City Road
London EC1V 1LR
Tel +44 20 7841 6900
Fax +44 20 7841 6910
E enquiries@laurenceking.com
www.laurenceking.com

Illustrations © Thibaud Hérem, 2020
© Mark Fletcher, 2020

A catalogue record for this book is available from the British Library.

ISBN 978-1-7862-7522-6

Illustrated by Thibaud Hérem
Designed by John Round Design
Contributing editor: Nick Yapp
Printed in China

Laurence King Publishing is committed to ethical and sustainable production. We are proud participants in The Book Chain Project®
bookchainproject.com

The Story of Trees

and how they changed the way we live

Kevin Hobbs & David West
Illustrated by Thibaud Hérem

Laurence King Publishing

Contents

Foreword

by Dr. Alexandra Wagstaffe
Crop Physiologist and Lecturer in Horticulture, Eden Project Learning

Like a meeting with old friends, this well-researched story of trees reminds us of trees we are familiar with while at the same time intriguing us with details and histories that are less or unfamiliar. Reading this book will provoke curiosity equally among well-seasoned plants people and those who are not, who may simply have stumbled across this impressive piece of work by chance. Nevertheless, there is no question about the value of trees and their contribution in supporting life on our planet as we know it.

The book maintains a delicate balance between fact-finding and storytelling. Each tree is introduced with fascinating botanical facts and examines geographical distribution over time, which seamlessly gives way to delightful anecdotes about how they have changed the way we live. Kevin Hobbs and David West treat the subject with a lightness of touch that belies the many extensive travels and fact-finding missions around the world. Comparable to modern-day plant hunters, their global awareness of the importance of plants to humankind is evident, told in charming, bite-sized ethnobotanical stories covering all continents. All the while, the narrative is set by the trees themselves, with the structure being guided by the chronological order in which they were evidenced to have played an important role in the lives of humankind.

Fascinating stories unfold, such as the English walnut (*Juglans regia*), 'one of humanity's oldest and healthiest foods', most recently celebrated for its high levels of Omega 3 fats and antioxidants. However, 'the earliest written reference to the walnut comes from the peoples of Chaldea in Mesopotamia, present-day Iraq. Inscriptions on ancient clay tablets proudly describe walnut groves in the Hanging Gardens of Babylon in about 2,000BC.' As a child within a European expat community, growing up in modern-day Iraq, historical references to the influence of Mesopotamia fascinate me, and I defy any reader to not find equally compelling stories of trees that have had a direct influence on them and their lives.

With 24 years' experience in crop physiology and plant sciences, my interest in trees has focused somewhat on physiological questions, such as how trees function. For example, for water to move from the roots to the uppermost leaves it has to travel a vertical distance of approximately 120 metres (393 feet) in some of the world's tallest trees, such as *Sequoia sempervirens*, (the California redwood). This truly monumental achievement is made possible by a complex interaction of anatomical adaptations, pathways and pressure differentials. However, it is not only the scientific complexity of trees that is astonishing, they also have the capacity to fill us with a sense of awe in their mere presence; for example, the wonderment we may feel as human beings when standing under trees that have witnessed their surroundings for hundreds and, in some cases, thousands of years.

The role trees play within cultures and communities can be explained by an innate human connection with the natural world, often described as 'Biophilia'. American naturalist Edward O. Wilson first defined this concept in 1984 as 'the urge to affiliate with other forms of life'. As such, this philosophy crosses the boundaries between scientific and spiritual exploration, which are successfully combined in *The Story of Trees*, where the authors have engagingly managed to convey the importance of trees while linking their traits to our human history throughout the ages.

Introduction

by Kevin Hobbs

Trees and the kingdom of plants are fundamental to our existence and development as a human race. However, apart from being aware of their contribution to the very air we breathe, few people realize the important role trees played in the lives of our ancestors and continue to play in our lives today – present as the paper on which these words are printed; enjoyed with a coffee from the coffee bean; or keeping us comfortable as a key material of our homes and their contents. We travel around in cars propelled by fossil fuel, on tyres made from rubber, along tree-lined streets. We fill our shopping baskets with numerous tree products, from fruit, nuts, herbs and spices to wine bottles sealed with cork. We decorate our homes using varnish that is a resin produced by a bug that feeds on trees, and clean our wooden-handled brushes with white spirit derived from the sap of a tree. The list goes on and on, encompassing medicine, toiletries, clothing and a great deal more. In fact, the relevance of trees in this modern, technological world continues, as demonstrated by the use of cork bark in the thermal insulation of spacecraft.

Our ancient ancestors, being closer to nature, used their native tree resources fully and often sustainably. For many, trees held great cultural or religious significance and, as such, were seen as a gift from, or representation of, deities. Reverence for the tree is evident throughout antiquity, and, indeed, persists in many cultures today. However, the wealth and ultimate success of many civilizations and countries was built on trade in their own resources, or those of others. Trade routes on land and sea contributed to the expansion of knowledge and the distribution of plant material. Where climate allowed, trees expanded beyond their original range and, often, became subject to natural or man-made hybridization.

Trees were among the first crops to be cultivated as humankind sought to take control of their resources. Unfortunately, the opposite was often the case for trees prized for their wood, or a derivative. They were obtained destructively, their very existence threatened by over-harvesting. Inevitably, civilizations or nations sought to take control of such valuable commodities. Trees and their products were at the centre of many conflicts, often at the expense of indigenous peoples. Great cruelty, as well as slavery, was endured as powerful nations and businesses vied for control. In many cases, trees were farmed in colonies far from their origin, ensuring stabilization of the markets, but often at the cost of ecological balance.

Today, international demand for products derived from trees continues to set commerce at odds with care for the environment. Even in these supposedly enlightened times, we continue to make mistakes in the way we manage farmed and natural trees. Invasiveness, spread of pests and disease and outbreaks of forest fires are among the problems we face. But from a more positive perspective, the appropriate use of trees in our changing climate brings great opportunity and benefit. There are many examples of success in managing biodiversity, bringing with it sustainable income, especially in regions where enterprise is limited.

Modern science continues to reveal fascinating details of plants and trees, both past and present. The earliest known tree grew from spores and belonged to a group of primitive fern-like plants called *Wattieza*, which has been dated to 385 million years ago. Palaeobotanists have reconstructed this long-extinct tree as being 9 metres (30 feet) tall, and palm-like in appearance. Pre-dating the dinosaurs by 140 million years, the forests of *Wattieza* and other primitive land plants removed carbon dioxide from the atmosphere, creating suitable conditions for the evolution of terrestrial animals and insects. Further fascinating insights into plant physiology have revealed more examples of the sophistication of trees, from detecting insect infestation in a neighbour to employing biochemical warfare on potential competition. We still have a great deal to discover about such capabilities.

Of the world's estimated 391,000 species of vascular plant, just over one quarter are considered trees. Many are known but little studied, and there are without doubt many more to discover. We must ensure that the world's natural habitats are conserved for future generations. A perfect example of a relatively recent discovery is the Wollemi pine, *Wollemia nobilis*, found just 150 kilometres (93 miles) from Sydney, Australia, in 1994 and identified as a new genus related to the monkey puzzle. Growing in just two stands in a narrow canyon were fewer than 40 trees. During the year that followed, successful propagation in tissue-culture laboratories led to this special conifer being grown in gardens in mild temperate regions, or warmer, throughout the world.

Within these pages, we hope to inform and inspire those who already have a love of trees, as well as those who otherwise may have taken them for granted. *The Story of Trees* is our story, but also that of our ancestors. It is about our relationship with some of the world's most important trees, both on a local scale and globally. With so many trees to choose from, we have endeavoured to feature those that have been, and in most cases continue to be, of cultural and practical value to humankind. Trees of every continent are included, with the exception of Antarctica, whose last trees were lost as the deep freeze took hold some three to five million years ago. These wonderful trees are presented in approximate chronological order as to humankind's first significant interaction with each one, and the book takes the reader from the boxwood implements made by the Neanderthals 171,000 years ago to the persimmon used for golf clubs in the nineteenth century. Height, speed of growth and longevity is given for each tree, and in each case this reflects the wild or natural conditions of its native range, allowing for variables such as altitude and soil type.

Today, a higher percentage of the world's population live in cities, bringing a much greater need for urban greening. Trees, and plants in general, not only look beautiful, but also have a positive impact on social behaviour and mental wellbeing. They improve biodiversity, and their leafy canopies help to cool and shade the otherwise sun-baked tarmac, all the time improving air quality.

Greater knowledge, respect and care for trees begins outside our front doors, while the story of trees continues, inextricably linked to our own.

Wollemi pine
(*Wollemia nobilis*)

Ginkgo biloba (Ginkgoaceae)

MAIDENHAIR TREE

Bearer of Hope

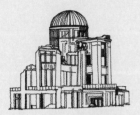

The ginkgo is a truly remarkable, deciduous conifer-like tree that earns its description as 'raining gold' from the spectacular coloured leaves that fall in autumn. It is, above all, a survivor, the only representative of an ancient order essentially unchanged for 200 million years. The ginkgo is effectively a 'living fossil', linking the present to the days of the dinosaurs of the Mesozoic era.

Although it is under threat as endangered in the wild, the ginkgo still grows wild on Mount Tianmu in China's Zhejiang province and is widely cultivated. In the temple gardens of Buddhism, Confucianism and Shintoism, it is considered a sacred tree, and some ginkgos are reputed to be more than 1,000 years old. An unmatched example of its resilience comes from more recent times. In the aftermath of the destruction of Hiroshima by the atomic bomb, at least six ginkgo trees soon began to reshoot only 1 kilometre (just over ½ mile) from the epicentre of the explosion. This seeming miracle of nature enhanced the tree's sacred status in Japan, where it became known as the 'bearer of hope'.

The ginkgo produces a remarkable, buff-yellow cherry-like fruit, which ripens on the female tree in autumn. It is notorious for its slimy, foul-smelling flesh. The fruit is harvested and soaked in water, then the flesh is removed leaving a white shell, which, once dried, is cracked open to reveal an edible green kernel. In Japan, these kernels are eaten roasted, salted or sweetened. They are believed to counteract the effects of alcohol, and are often served in bars. However, they should be consumed only in small amounts, as they contain harmful neurotoxins.

The tree was first planted in Europe in the eighteenth century as an ornamental, and became a popular street tree. To avoid the foul-smelling fruit, only male trees were cultivated, but female trees did somehow join them, and the consequences were understandably unpopular with local residents.

OTHER COMMON NAME
Ginkgo tree

ORIGIN
Zhejiang (China)

CLIMATE AND HABITAT
Moist, warm temperate regions in
a range of reasonably fertile soils

LONGEVITY
At least 1,000 years

SPEED OF GROWTH
30–50 centimetres/
12–20 inches per year

MAXIMUM HEIGHT
35 metres/115 feet

*In China, the gingko nut is known as
the 'silver apricot' and is often eaten at
weddings as it is considered auspicious.*

OTHER COMMON NAMES
Common yew, European yew

ORIGIN
Europe, western Asia, North Africa

CLIMATE AND HABITAT
Tolerates a wide range of soils in
a temperate or subtropical climate;
most commonly on limestone in
dense or sparse mixed forest

LONGEVITY
500 years; some examples
are believed to be more
than 5,000 years old

*The bright red flesh of the
yew seed is the only part of
the tree that is not poisonous.*

SPEED OF GROWTH
10–20 centimetres/
4–8 inches per year

MAXIMUM HEIGHT
20 metres/66 feet

Taxus baccata (Taxaceae)

YEW

Guardian of the Cemetery

The yew, a tree of mystery and legend, is one of Europe's longest-living species. It is a tree full of character, and each is an individual, with beautifully twisted fissured branches, trunks and, often, exposed roots.

A specimen in the churchyard of St Cynog, Defynnog, in Powys, Wales, has been dated as 5,063 years old. It was a seedling in about 3045 BC, and began life 500 years before the Great Pyramid of Giza was built. The oldest yew artefact is 400,000 years old. Known as the Clacton Spear, it was recovered from Palaeolithic sediment at the seaside town northeast of London, and is currently the earliest example of a worked wooden implement in the world.

To ancient Mediterranean civilizations – Egyptian, Roman and Greek – the yew was a symbol of death. In Shakespeare's *Richard II*, the tree is described as 'the double-fatal yew', most of its parts being poisonous and its wood fashioned for centuries into spears and bows. In the hands of a strong archer at the time of Agincourt, the yew was flexible and powerful enough to shoot an arrow well over 250 metres (820 feet). The need to keep England well armed led Henry IV to empower his agents to enter private land and cut down yews.

Yews abound in churchyards, many of them pre-dating the church buildings themselves, for early Christians often adopted pagan sites, symbols and festivals, whether Druidic or Celtic in origin. Devotees of the Christian Church often used yew branches in place of palms on Palm Sunday.

Sadly, the natural form of *Taxus baccata* is seldom planted in gardens or the wider landscape. Its many cultivars include those with golden foliage, as well as the popular Irish yew, 'Stricta' or 'Fastigiata', a neat, tight, upright grower, along with dwarf forms. This handsome evergreen conifer, used for hedging and topiary, deserves a place in every garden, however small.

Buxus sempervirens (Buxaceae)

COMMON BOX

The Art of Clipping

In 2012, on a building site in Tuscany, Italy, an incredible hoard of wooden and bone implements fashioned by Neanderthal man was unearthed. They were dated to the late middle Pleistocene age, some 171,000 years ago, when the region was inhabited by early Neanderthals. Along with the fossilized bones of *Palaeoloxodon antiquus*, a straight-tusked elephant, were found box branches worked into tools. At almost 1 metre (3 feet) long, these pieces of box had been sharpened at one end for digging with the other end rounded into a handle. Numerous cuts and grooves indicated the use of stone tools during its manufacture, and superficial charring was thought to be the result of using fire to complete the process, providing a rare insight into the Neanderthal intellect.

The choice of box was a good one. The tree is slow-growing and produces the toughest woods of its natural range. Box wood is so dense that it will not float in water; understandably, unlike *Hearts of Oak*, there is no song in praise of the box. Bees are attracted to the small creamy-white to yellowish-brown male and female flowers on each plant, although the flowers themselves have an unpleasant fragrance. They are mainly pollinated by the wind, however, and produce modest brown capsules containing black seeds. Box trees also provide a dense, sheltered habitat for small birds, mammals and insects.

Common box has been used in gardens as hedging and topiary since Roman times, and is still seen in formal gardens today. Parterre gardens became popular during the Renaissance, and their precise, symmetrical patterns are testament to the designers, and to those who maintain them. Early examples contained elaborate mazes, labyrinths, topiary of animals and birds, and knot gardens. Sadly, the box is used less often in topiary today, owing to the increasing prevalence of box blight, a fungal disease that produces unsightly bare patches and ultimately leads to the box's death.

OTHER COMMON NAMES
European box, boxwood

ORIGIN
Southern Europe, North and West
Africa; naturalized and possibly
wild in the south of England

CLIMATE AND HABITAT
An understorey plant to much
larger trees, or on open hillsides in
cool temperate to Mediterranean
climates; usually on chalky soil

LONGEVITY
150–200 years

SPEED OF GROWTH
5–15 centimetres/
2–6 inches per year

MAXIMUM HEIGHT
8 metres/26 feet

*All parts of the common box are
toxic, including the leathery
evergreen leaves.*

OTHER COMMON NAME
Common fig

ORIGIN
Southwestern Africa

CLIMATE AND HABITAT
Dry, sunny situations, often in
nutritionally poor soil in a warm
to hot subtropical climate

LONGEVITY
At least 200 years

SPEED OF GROWTH
20–50 centimetres/
8–20 inches per year

MAXIMUM HEIGHT
10 metres/33 feet

*Black figs have a high sugar content
and are sweeter than green figs.*

Ficus carica (Moraceae)

Fig

Pollinated by Wasps

The fig is probably the oldest cultivated fruit on earth. The deciduous fig tree grows rapidly, sending out – even from the base of the trunk – flexible branches along which the tough green leaves and bulbous fruits appear. The bark of the tree has a silvery sheen, and the fruit varies in colour according to the variety or cultivar. Figs may be green, brown or even a rich deep purple. The flowers of the fig are invisible, as they bloom inside the fruit itself. Figs growing in the wild are pollinated by the small black wasp *Blastophaga psenes*. The pollinator ultimately dies inside the fruit, where an enzyme called ficin breaks down the wasp into protein.

Fig trees have been cultivated widely throughout the world. They are loved by many and loathed by some – usually those with memories of the wartime laxative syrup of figs, which tasted nothing like fresh or dried figs. A relatively recent discovery by Israeli archaeologists on the shores of the Sea of Galilee shows evidence of fig trees being cultivated some 23,000 years ago. The inscriptions of King Urukagina, ruler of the city-state of Lagash, Mesopotamia, around 2400 BC and author of the first recorded legal code, contain a reference to the fig as a crucial source of food. To the Greeks and Romans, it was the gift of the gods. But in English, the word 'fig' has bad connotations. It has long been used to mean something of no importance, and the Shakespearean version of the present-day V sign was known as the 'fig of Spain'.

A flowering plant of the mulberry family, native to the Middle East and western Asia, the fig tree is said to like 'its feet in hell and its head in heaven'. It can grow to a height of 10 metres (33 feet) and a similar width, while its roots will thrive in restricted space and even in piles of rubble with little earth.

The fig is notorious for its mention early in the Bible (Genesis 3, verse 7). After they had eaten the forbidden fruit, the eyes of Adam and Eve were opened 'and they knew that they were naked; and they sewed fig leaves together, and made themselves aprons'. They made the right choice: fig leaves are large and resilient, but the sap of the green part of the fig itself is an irritant to human skin.

Eucalyptus globulus (Myrtaceae)

TASMANIAN BLUE GUM

Music from the Ashes

The eucalyptus is synonymous with Australia, but in fact the oldest fossil records for this tree – dating back more than 50 million years ago – have been found in Patagonia, Argentina. Perhaps this apparent discrepancy is not surprising, since geologists believe that Australia and South America may have been physically linked to each other by Antarctica at that time. The ancient fossils resemble many of the 900-strong eucalyptus species of today.

The Tasmanian blue gum was first studied by Jacques de Labillardière, a French botanist who visited Tasmania in 1792. A man who was said to 'hide everything that was good in his soul behind a caustic intellect', de Labillardière was a friend of the famed English naturalist Sir Joseph Banks, and he toured the world studying such specimens as the Tasmanian blue gum. This species thrives in Australia and Tasmania because of its ability to adapt to fire. Eucalyptus oil is flammable, so the shed bark and fallen leaves of the tree serve as ideal fuel for a forest fire. Once the fire has burnt itself out, however, with most competition eliminated, the tree is reborn from dormant buds beneath its charred bark. The seed capsules open in the heat, and the ash-rich soil is perfect for germination.

Early European settlers (voluntary or coerced) made use of the evergreen Tasmanian blue gum for its durability and toughness. Sadly, commercial farming of the blue gum is believed to have subsequently disrupted the ecological balance in Australia, where for thousands of years the indigenous people had managed the land, creating grasslands that both attracted animals they could hunt and which provided natural firebreaks. Today the tree is still used by Aboriginal peoples: the bark, when dried and flattened, as canvas; the stems, when hollowed out by termites, as didgeridoos.

The Tasmanian blue gum, like all eucalyptus, is a fast-growing tree, adding several metres a year to its height, and has many uses when harvested. The timber itself is still used in bridge construction, as flooring and other surfaces subject to hard wear. The young leaves, which are powder blue, give their name to the tree; the mature green leaves and the flowers have long been the principal source of eucalyptus oil, which is used medicinally to bring comfort and relief from blocked noses.

OTHER COMMON NAMES
Blue gum, southern blue gum

ORIGIN
Tasmania and southern
Victoria (Australia)

CLIMATE AND HABITAT
Tall, open forests on a wide
range of neutral to acid soils in
a subtropical or temperate climate

LONGEVITY
Up to 200 years

SPEED OF GROWTH
2–3.5 metres/6–12 feet per year

MAXIMUM HEIGHT
70 metres/230 feet

*Flowers from the Tasmanian blue
gum are loved by honey bees and
produce a strong-tasting honey.*

OTHER COMMON NAMES
Great Basin bristlecone pine,
intermountain bristlecone pine

ORIGIN
The mountains of Utah,
Nevada and eastern California
(United States)

CLIMATE AND HABITAT
Full sun, low rainfall situations
with extremes of temperature from
-18°C to 34°C/0°F–93°F. Open,
rocky situations at high altitude;
denser mixed forests at low altitude

LONGEVITY
300 years at low altitudes; 1,000
years at high altitudes. Some
examples are believed to be
more than 5,000 years old

SPEED OF GROWTH
10–50 centimetres/4–20 inches per
year, age and conditions permitting

MAXIMUM HEIGHT
16 metres/52 feet

*Female cones take two years to
develop, and start off a deep purple.*

Pinus longaeva (Pinaceae)

Bristlecone Pine

Sculpted by Time

Of the world's oldest living trees, few come close to matching the character and gnarled beauty of the bristlecone pine. With its extraordinary appearance come extraordinary properties. In the United States there is a dead bristlecone pine, reckoned to have germinated 9,000 years ago, but still standing – short, stout, gnarled and bleached silvery white by the cold winds of high altitude. In the White Mountains of California there is a living bristlecone pine that is believed to be 5,068 years old, although some have disputed this figure. It is difficult to be sure, because the exact location of the tree has been kept secret, but if that were accurate, this would be the oldest known single tree of any species.

The bristlecone pines of the Great Basin in the western United States were first joined by early Paleo-Indian hunters towards the end of the last great ice age, around 12,000 BC. Much later, these early inhabitants established settlements to become Indian tribespeople such as the Paiute. Living in harmony with the land, they harvested the edible nuts of the piñon pine (*Pinus edulis*), and favoured the strong wood of the bristlecone for building shelters.

The great age and characteristically sculpted appearance of the bristlecone pine are born of the harsh environment to which it has adapted. Isolated groves and lone trees grow extremely slowly, and in some years do not even produce a growth ring. This results in very dense and hard wood that reduces the tree's vulnerability to pests, diseases and harsh weather. Even its green needles are long-lived, and some stay on the tree for up to 40 years before dropping to the ground. The female cones, from which the tree takes its name, begin life dark purple, with inward-curled bristles on the surface, but as they mature they change to brown.

The bristlecone pine's great age has proved to be of huge benefit to modern scientific research. Its annual growth rings have helped scientists to calibrate carbon-dating techniques, and provide weather data going back thousands of years, contributing to the study of our fluctuating climate.

Pinus pinea (Pinaceae)

STONE PINE

The Dancer's Friend

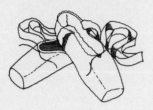

Fans of epic Hollywood films set in ancient Rome will be familiar with the Appian Way. Stone pines still line this approach road to Rome, standing straight and tall, with their distinctive dark, spreading, umbrella-shaped canopies that have provided shade for thousands of years. The stone pine is at the centre of an image that, like the Colosseum, is the essential Rome.

In fact, the stone pine is far older even than ancient Rome, and dates back to Neolithic times. Stone-pine cones have been found among offerings made as part of the funeral rites of the Egyptian Twelfth Dynasty. In early, and more humid, times the stone pine grew in the Sahara Desert. Its nuts were distributed along the trade routes of ancient Mediterranean cultures, where the tree itself had cultural and spiritual significance. Today the stone pine can be seen in every country that borders the Mediterranean, as well being grown as an ornamental around the world in mild, temperate climates.

The timber of the stone pine is too coarse and resinous to be of great commercial value, although it serves well enough in furniture-making. Most profitable is the resin itself. This, like rubber, is tapped from the trunk of the tree and used as waterproofing or varnish. In its hard form or as a powder, the resin has been used for centuries to lubricate the bows of stringed musical instruments (from violins to double basses), and to prevent dancers' shoes from slipping in ballet and Irish dancing.

The seeds inside the pine cone ripen in the third year and are released in response to heat – the thick bark of mature trees is fire resistant. These seeds, or 'pine nuts', are one of the three essential ingredients of the popular sauce pesto (with basil and Parmesan cheese), and since antiquity they have been the traditional accompaniment to meat, fish and salad dishes in Mediterranean and Middle Eastern cuisine.

OTHER COMMON NAMES
Umbrella pine, parasol pine

ORIGIN
Mediterranean

CLIMATE AND HABITAT
Mediterranean: long, dry summers
and low humidity; maritime
sands and alluvial soils

LONGEVITY
Up to 300 years

SPEED OF GROWTH
30–50 centimetres/
12–20 inches per year

MAXIMUM HEIGHT
25 metres/82 feet

Stone-pine cones take three years to
mature, longer than any other pine.

OTHER COMMON NAMES
Alligator pear, butter
fruit, summer pear

ORIGIN
South Central America

CLIMATE AND HABITAT
Subtropical climate, frost-
free, humid; fertile, deep,
well-drained soils

LONGEVITY
At least 100 years

SPEED OF GROWTH
80–100 centimetres/
32–40 inches per year

MAXIMUM HEIGHT
20 metres/66 feet

*The avocado, like the banana, is
a climacteric fruit, which means
that it matures on the tree, but
continues to ripen once picked.*

Persea americana (Lauraceae)

Avocado

Food for a Prehistoric Giant

From its home of Mexico and Central America, the increasingly popular superfood avocado has a long and fascinating history. Botanically a drupe, avocado is a fleshy fruit that should, more accurately, be described as a single seeded berry.

The avocado is a handsome, spreading tree. It grows to a maximum height of 20 metres (66 feet), thickly covered in leaves that in colour and shape resemble bay leaves. The fruit form at the end of its branches, hanging in clusters and weighing the branches down. The avocado tree's survival once relied on its being eaten whole by animals, which then distributed the seed. The giant ground sloth, *Megatherium*, was believed to be such an animal, but has been extinct for some 11,000 years, so the avocado has had to rely on smaller mammals since then, including squirrels and man. Archaeologists have discovered that avocados were eaten by the inhabitants of Huaca Prieta, Peru, the oldest known human settlement in the Americas, dating back 15,000 years.

The name 'avocado' is derived from *āhuacatl*, the Mexican Nahuatl word for testicle, because of the shape of the avocado, the fact that avocados often grow in pairs, and because the Nahuatl believed that eating the avocado enhanced fertility. The milky juice of the seed is high in tannin and darkens when it is exposed to the air, and it was found to be useful as an alternative to ink by the Spanish conquistadors. Examples of historic documents written in distinctive red-brown avocado ink can be found in the archives of the city of Popayán, Colombia, to this day.

The avocado also yields an oil that is used in creams and cosmetics, and is favoured over olive oil because it is more easily absorbed by the skin. It also makes a fine ornamental plant, and those of us who place avocado seeds in small pots of water are always delighted when a seedling sprouts.

Prunus persica (Rosaceae)

PEACH

A Roman Recipe

En masse, few trees can match the beauty of an orchard of peach trees. The children's author Roald Dahl had originally intended to write a book about a giant cherry, but the book became *James and the Giant Peach* because, said Dahl, 'a peach is prettier, bigger and squishier than a cherry'.

The Latin name, *Prunus persica*, wrongly identifies the peach's origin as Persia, which is now known to have been a stopping point on its journey from the Far East. Recent archaeological finds dating back 2.5 million years have established that the fruit's (or, to be botanically exact, drupe's) true origin was China. At the Xishuangbanna Tropical Botanical Garden in Yunnan province, fossilized peach stones have been unearthed that pre-date *Homo erectus* by at least 700,000 years. The peach moved on to Greece and Rome, and became widely cultivated throughout the Mediterranean, where it was eaten both fresh and pickled. Still heading west, with early Spanish explorers, the peach reached the Americas, becoming naturalized in Mexico and the southern United States. There, where they were known as 'Tennessee naturals', peaches were turned into a sweet wine.

Historically, the peach has sexual connotations. It features with other fruit in Byzantine, Gothic and Renaissance paintings, its presence full of symbolism aimed at the morals and beliefs of the masses. A ripe peach represented virtue and honour, whereas a half-eaten or, even worse, rotten peach denoted a woman who had damaged her reputation through immoral or shocking behaviour.

The peach has its place in literature and folklore. Apicius included it in his cookbook *De re coquinaria* (On the Subject of Cooking), and Pliny the Elder mentioned the peach in his voluminous *Naturalis historia*. An Italian proverb tells us to 'Give a fig to a friend, a peach to an enemy'; a Vietnamese proverb has a variation on this, 'Receive a plum, return a peach'.

In Victorian England, the gardeners of wealthy landowners developed inventive ways of growing fruit-bearing peach trees, usually against sunny walls or in heated conservatories. Both produced good results, but at considerable expense.

OTHER COMMON NAME
Common peach

ORIGIN
Northwestern China

CLIMATE AND HABITAT
Man-made or disturbed habitats;
forests and forest edges in a warm
or cool climate; well-drained, acidic
to neutral loam or sandy soils

LONGEVITY
15–25 years

SPEED OF GROWTH
50–100 centimetres/
20–40 inches per year

MAXIMUM HEIGHT
6 metres/20 feet

*White-fleshed peaches, popular in Asia,
are less acidic than yellow-fleshed ones.*

OTHER COMMON NAME
European olive

ORIGIN
Mediterranean

CLIMATE AND HABITAT
Poor, dry soil, ideally with access
to deep moisture; full sun in a mild
or warm temperate climate

LONGEVITY
At least 1,000 years; some are
claimed to be almost 2,000 years old

SPEED OF GROWTH
10–30 centimetres/
4–12 inches per year

MAXIMUM HEIGHT
15 metres/50 feet

*The difference between green
and black olives is ripeness –
black olives being ripe.*

Olea europaea (Oleaceae)

OLIVE

Athena's Gift

The olive tree, its fruit, and the oil that is distilled from it are considered quintessentially Mediterranean. Cultivated since prehistoric times, the olive was first used by humans at least 20,000 years ago. Over the centuries many Mediterranean countries built their wealth on trading in olives. An archaeological site on the southwestern shore of the Sea of Galilee revealed olive stones and wood fragments left behind by Upper Palaeolithic hunter-gatherers. Pottery vessels found nearby provided early evidence that olive oil was traded some 8,000 years ago.

Exactly when olive farming began is much contested, but written evidence from 2400 BC has been found on Syrian clay tablets. A fire that destroyed the city of Ebla, on the outskirts of Aleppo, baked the clay tablets, preserving the writing. Most dealt with commercial affairs, and the tablets revealed that the value of olive oil was put at five times that of wine. In modern Palestine, ownership of an olive tree is much treasured. Over the centuries, single trees were bequeathed to more than one descendant, and the division of the tree has continued so that now each branch of the tree is owned by a different member of the family.

The ability of the tough, wrinkled olive tree to thrive in poor, dry soil allowed it to spread across the Mediterranean and beyond. In Greek mythology it is said that the goddess Athena, daughter of the supreme god Zeus, gave the olive tree to the Athenians. The tree is believed to have appeared next to the well of the Athenian Acropolis, becoming the source of all future trees. A wreath of olive leaves found in the 3,300-year-old sarcophagus of the Egyptian king Tutankhamun is now preserved in the herbarium of the Royal Botanic Gardens at Kew.

The olive branch is still seen as a symbol of peace. In the Bible it denotes light, peace and divine blessing. References to olives and olive trees are found throughout the Qur'an, and Muslim prayer beads are made of olive wood. When defeated in battle, Greeks and Romans held up olive branches as a symbol of surrender.

Corylus avellana (Betulaceae)

HAZEL

Of Mystical Powers

Hazel trees come in many forms: American, Asian, Chinese, Turkish and Himalayan. The hazel is best known for the edible nuts that each of these species produces. Strictly speaking, the nuts from cultivated hazels are called filberts. As well as being a nutritious food, the hazelnut was used in medicine throughout antiquity. The ancient Greeks extolled its healing qualities for conditions ranging from the common cold to baldness.

Often more a bush than a tree in appearance, in winter the hazel produces rather hairy twigs on which there are round green buds. The deep green leaves, which appear in April, are almost circular, with toothed edges, crowned with a large tooth at their tip. In late winter the hazel can be recognized by its long, golden yellow male catkins, which hang like lambs' tails from the long, thin stems. The female catkins are far harder to find – tiny green buds scattered along the same stems. In the summer, the female flowers turn into nuts that ripen in the autumn.

The hazel is an enthusiastic tree with considerable powers of regeneration when cut back. Of its own volition, it sends out shoots from near the ground, and it reacts with similar swiftness to the process of coppicing (from the French *couper* – to cut). This has made the hazel a boon to human beings. For centuries it has been used as the backbone of hedges, as the wattle (wooden framework) of wattle-and-daub walls in cottages, as the skeleton of thatched roofs and to make excellent bean poles.

An air of mystery surrounds the hazel. In the nineteenth century, the Brothers Grimm wrote a story in which the hazel hides the Christ Child's mother from a voracious adder: 'As the hazel has been my protection,' said Mary, 'it shall in future protect others also'. For hundreds of years cuttings from hazel trees have been used by dowsers to find sources of water, a feature that Arthur Ransome made great use of in *Pigeon Post*, one of his Swallows and Amazons novels, in which Titty manages to hold a springy fork of hazel between her thumbs and forefingers to use as a divining rod to detect the existence of an underground spring during a Lake District drought.

OTHER COMMON NAMES
Cobnut, European hazel, aveline

ORIGIN
From British Isles eastwards
to Russia and the Caucasus,
and from central Scandinavia
southwards to Turkey

CLIMATE AND HABITAT
Moist soils of lowland woods
and forests; also found growing
in hedges, meadows and along
the banks of streams, as well as
on wasteland. Tolerant of a wide
temperature range and dry periods

LONGEVITY
At least 80 years (when coppiced)

SPEED OF GROWTH
45–100 centimetres/
18–40 inches per year

MAXIMUM HEIGHT
15 metres/50 feet

*The kernel, or middle,
of the hazelnut is the basis of praline.*

OTHER COMMON NAME
Chian turpentine

ORIGIN
Asia Minor, Mediterranean

CLIMATE AND HABITAT
Dry, open woods and
scrubland close to sea level
in a subtropical climate; full
sun; often limestone soil

LONGEVITY
Up to 500 years

SPEED OF GROWTH
10–20 centimetres/
4–8 inches per year

MAXIMUM HEIGHT
10 metres/33 feet

*In spring the terebinth produces
maroon-red flowers, from which
red fruit and then nuts form.*

Pistacia terebinthus (Anacardiaceae)

TEREBINTH

Tree of the Mycenaeans

The terebinth deserves our gratitude for producing a glorious display of colour in the depths of winter. The bare stems of the tree are suddenly brightened by a simultaneous flush of new leaves and maroon-red flowers. From these flowers come clusters of small, rounded red drupes, roughly the size of a pea, that ripen to black. The nut itself is very palatable, sweeter and oilier than the almond.

This small deciduous tree belongs to the cashew family. Its mature height is around 10 metres (33 feet), and its leaves are long and leathery, reminiscent of the leaves of the carob tree. Each leaf is made up of between five and eleven glossy leaflets, coloured bright green. Through its resin and oil, the entire tree emits a strong and bitter smell. The trees are prey to galls, excrescences caused by insects. These galls are shaped like the horns of a goat, which is why the terebinth's common name in Spanish is *cornicabra*. Despite the presence of these galls, the terebinth is strong and resilient, growing in mountain areas where few other species survive.

Commercially, the terebinth produces turpentine. Resin is collected from incisions made in its bark, and is then distilled. The antibacterial properties of turps have long been exploited by man, and its presence in wine residue has been found in 7,000-year-old jars excavated from an archaeological site in the Zagros Mountains of Iran.

The first written record of the tree comes from 3,500-year-old Linear B clay tablets from Mycenae. Later there were many references to the terebinth in the Bible, the most significant of which is to be found in the First Book of Samuel, Chapter 17. The Valley of the Terebinth (the Valley of E-lah in the Bible) is said to be the setting for David's slingshot victory over Goliath. However, it is now believed that the elah tree of the Bible is actually *Pistacia palaestina*, a species with similar properties and appearance.

Juglans regia (Juglandaceae)

ENGLISH WALNUT

In the Hanging Gardens
of Babylon

Like the hazel, the English walnut is generous in its gifts. Ornamentally it is handsome, with a smooth olive-brown bark that turns silver-grey as the tree matures, large green compound leaves (composed of a main stem with many separate leaflets), drooping male catkins, and female flowers in clusters of two to five, which become the green-husked fruit. This darkens as the nut inside ripens. It is one of humanity's oldest and healthiest foods – a quarter of a cup of walnuts contains 100 per cent of the recommended daily supply of Omega 3 fats and antioxidants. Eating walnuts helps people with heart disease, may reduce the risk of prostate and breast cancer, is good for the brain and can counteract Type 2 diabetes. Scientists have also discovered that walnuts contain melatonin, which regulates the sleep/wake cycle of humans and animals.

Widely cultivated in temperate regions, the walnut is native to the mountain valleys of Central Asia, where it is known to have been grown by Neolithic people some 7,000 years ago. Like many native Asian trees, it worked its way westwards from China to the Caucasus, Persia, Greece and Rome. The Romans introduced it to England, from where English merchant sailors later transported it around the trade ports. From that habit came another common name, the English walnut.

The earliest written reference to the walnut comes from the peoples of Chaldea in Mesopotamia, present-day Iraq. Inscriptions on ancient clay tablets proudly describe walnut groves in the Hanging Gardens of Babylon in about 2,000 BC. In the Song of Solomon, Chapter 6, Verse 11, Solomon is thought to have been referring to the walnut when he said 'I went down to the garden of nuts to see the fruits of the valley...'

The wood of the walnut is firm and smooth enough to be used in sculpture and carving, and hard enough to be used for the stock of sporting guns. Timber from the walnut tree has for centuries been turned into elegant and, it has to be said, costly tables, chests and bureaux. The tree also provides food for leaf-eating moth caterpillars and is a tasty diet for mice and squirrels.

OTHER COMMON NAMES
Persian walnut, common walnut

ORIGIN
Kyrgyzstan, Tajikistan,
Uzbekistan, Turkmenistan

CLIMATE AND HABITAT
Moist, deep soil in sunny situations
in variable temperate climates,
typically in small stands

LONGEVITY
At least 70 years

SPEED OF GROWTH
20–40 centimetres/
8–16 inches per year

MAXIMUM HEIGHT
35 metres/115 feet

*For a long time the walnut was
thought to be the seed of a drupe, but
now it is widely described as a nut.*

OTHER COMMON NAMES
Terebinth nut, common pistache

ORIGIN
Turkey, Iran, Syria, Lebanon,
Southern Russia, Afghanistan

CLIMATE AND HABITAT
Hot, semi-arid regions, often
in poor soil; tolerant of salt

LONGEVITY
Up to 150 years

SPEED OF GROWTH
10–60 centimetres/
4–24 inches per year

MAXIMUM HEIGHT
10 metres/33 feet

*Every two years the pistachio tree
produces about 50,000 seeds.*

Pistacia vera (Anacardiaceae)

PISTACHIO

Sweet and Green

The pistachio is a tough tree, brought up in arid, desert-like conditions throughout the Middle East. It can withstand temperatures ranging from -10 to 48°C. Like its relative the terebinth (see page 34), it has a short trunk topped with a large canopy of leaves. It is not without its season of beauty, the autumn, when the clusters of ripening drupes, as thick as bunches of grapes, change from green to a stunning yellow and red. When fully ripe, the shells of the pistachio split open with an audible pop. Persian legend holds that lovers who meet in pistachio orchards on moonlit nights will receive good fortune if they hear the shells crack open.

The seed of the pistachio is not a botanical nut, but a culinary one, and has been recognized as important for more than 9,000 years. Like the terebinth, the pistachio grew up in the Zagros Mountains of Iran – where pistachio remains dating to 6750 BC have been found – and again like its cousin, the pistachio has its place in legend. The Queen of Sheba is said to have demanded that the entire pistachio harvest from her lands be reserved for her and her court alone. In the Qur'an, the food is said to be one of those brought by Adam from Paradise to Earth.

Well over 50 varieties of pistachio have been cultivated in the last 3,000 years. The popularity of the pistachio as a food – whether eaten on its own, as an ingredient in salads, served with meat or fish, or as an ice cream – has grown enormously in modern times.

In the late nineteenth century pistachios exported from the Middle East were often dyed red. There is debate as to the reason. The popular theory relates to a Syrian dealer named Zaloom, who did it to distinguish his pistachios from those of his rivals. Another, more prosaic theory is that pistachios were dyed to cover blemishes and stains on their shells. Whatever its origin, the practice disappeared after the establishment in the 1970s of large-scale production in the United States.

Toxicodendron vernicifluum (Anacardiaceae)

LACQUER TREE

Toxic Treasure

The Chinese lacquer tree, also known as the Japanese lacquer tree, is cultivated for the sap it produces when tapped. A mature tree may grow to 20 metres (66 feet) in height, producing large green compound leaves composed of many leaflets, similar to those of the common ash and rowan tree. In spring, when blossom covers much of the tree, and in autumn, when the leaves turn a brilliant red, the tree has great beauty. Its fruit, which resembles tubby figs, and its leaves are used in Chinese medicine to treat internal parasites and to stop bleeding.

The tree's great commercial value, however, comes from its resin. When the lacquer tree is ten years old, its trunk is cut open in rows of deep horizontal incisions, and small basins are fixed to the trunk below to collect the greyish-yellow resin. This tapping takes place every five days, and the incision turns black when its supply of resin is exhausted. It needs care and skill on the part of the cutter, for the resin contains the allergenic compound *urushiol*, a substance found also in poison ivy; even the vapour from the resin can cause a skin rash.

After being harvested and allowed to mature, the resin becomes the varnish that has been used for centuries on wooden furniture, utensils, carvings and statues. In modern times Korean, Japanese and Chinese users firmly believe that this varnish or lacquer is far superior to synthetic substitutes. It has certainly stood the test of time. One red lacquered wooden bowl unearthed in the Hemudu Ruins in Yuyao, Zhejiang Province, China, has been dated to sometime between 5,000 and 4,000 BC.

A bizarre use of the resin was practised in northern Honshu, Japan between the eleventh and twentieth centuries, when a small school of Buddhist monks carried out a process of 'live' self-mummification. After spending six years on a special diet, a monk would drink *urushi* (lacquer) tea that poisoned, but ultimately preserved, his body. Sealed in a tomb with a breathing pipe, the monk sat in the lotus position until he died, at which point the tomb was closed for 1,000 days before the body was removed. Full mummification was considered successful if the monk's body was still in the lotus position, after which his remains were considered sacred.

OTHER COMMON NAMES
Chinese lacquer tree, Japanese
lacquer tree, varnish tree

ORIGIN
China, Japan

CLIMATE AND HABITAT
Woods and thickets on mountain
slopes in a cool to warm temperate
climate; well-drained fertile soils

LONGEVITY
At least 60 years

SPEED OF GROWTH
30–60 centimetres/
12–24 inches per year

MAXIMUM HEIGHT
20 metres/66 feet

*The leaves of the lacquer tree are high
in tannin, and when they fall in autumn
are sometimes used as a brown dye.*

Other common name
Date

Origin
North Africa, Arabian peninsula

Climate and habitat
Open, sunny sites in warm
temperate to dry tropical
regions; free-draining soil
with access to moisture

Longevity
Up to 100 years

Speed of growth
20–30 centimetres/
8–12 inches per year

Maximum height
25 metres/82 feet

*A single bunch of dates can
contain as many as 1,000 fruit.*

Phoenix dactylifera (Arecaceae)

DATE PALM

Food for Masada

The date palm has been a valuable source of food for many civilizations throughout its native lands in the Middle East and North Africa. To the ancient Egyptians it was more than a sweet and nourishing fruit. The tree permeated their culture. Its presence, even in the driest of deserts, was seen as a miracle of fertility and life, and the ray-like arrangement of its leaves symbolized those of the Egyptian sun god Ra. Ancient Greek and Roman cultures also associated it with the sun in their architecture and even in the design of their coins. In Hebrew and Christian cultures, the date palm has long been seen as a symbol of peace. In modern times, however, the date palms that line the waterfronts of Mediterranean cities, from Tripoli to Tel Aviv and from Marbella to Majorca, are more likely to be seen as decorative symbols of wealth and luxury.

The handsome trunk of the date palm is clad from top to bottom with overlapping woody leaf bases, or boots, from old leaves that have died. The trees are topped by the live leaves or palms, long and feathery, which give shelter to the dangling clusters of the date fruit. Staggeringly, a single mature tree can produce between 70 and 140 kilos (150–310 pounds) of dates in one year.

In 2005 the date palm provided an incredible link to a historical moment of great significance. In AD 73, after a siege that had lasted some months, the walls of the city of Masada (in present-day Israel) were breached. Jewish Zealot defenders were ordered to surrender to the Roman forces. The Zealots chose suicide rather than capture. Some 1,932 years later archaeologists discovered remains of their food reserves, among which were dates. Researchers sowed one of the date stones and were amazed when it germinated. It is, so far, the oldest germinated tree seed, and the seedling has given Israeli scientists an extraordinary opportunity to study the genetic relationships between ancient and modern date palms.

Fraxinus excelsior (Oleaceae)

COMMON ASH

Of Norse Mythology

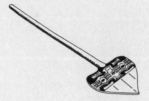

The common ash is a tree that seems to leaf grudgingly in late spring and then, miser-like, hold on to those leaves until late autumn. The tree is common by name and by nature. It is found throughout Europe, from the Arctic Circle to Turkey, thriving even where there are limestone rocks, for the ash can tolerate calcium in the soil. Its name is derived from both the Norse *askr* and the Anglo-Saxon *aesc*.

Although it is often used in hedging, if it is given enough space in which to grow the ash can become large and magnificent, its bare winter stems easily identified by the distinctive black buds. Its bark is appropriately ash-grey, smooth at first but maturing into a regular network of raised ribs, like a fishing net. It can grow to 35 metres (115 feet) in height, but its girth is narrow, normally about 4.5 metres (15 feet). Its life span is rarely more than 150 years, although it can live up to 400 years.

The wood of the ash is strong and resilient, and that, along with its abundance and rapid growth, continues to ensure its use in woodworking, and as firewood or charcoal. Its ability to absorb impact without splintering makes it valuable for the handles of garden tools such as forks, spades and axes, for picks and hammers, and for sporting equipment such as hockey sticks.

In Norse mythology, the ash holds a high place. Yggdrasil, the Norse 'tree of life', was the eternal ash tree that grew at the centre of the nine worlds. Rooted in paganism that persisted beyond the arrival of Christianity, the ash features in the chronicles of such deities as Thor, Odin and Freya. Painted and decorated ash paddles found in Horsens Fjord in Jutland, Denmark, have been carbon-dated to about 4,700 BC. They would have been used to propel the dugout boats of hunter-gatherers during the late Mesolithic Age, from 5,300 to 3,950 BC.

The tree is a very effective colonizer, and can establish itself on land left open by fallen neighbours. This ability is not appreciated by gardeners, however, who see the plant as a weed. Sadly, the ash's abundance is now under threat from ash dieback. This infection comes from *Hymenoscyphus fraxineus*, a fungus powerful enough to lead to reports in 2016 that the ash tree was in danger of extinction in Europe.

OTHER COMMON NAMES
Ash, European ash

ORIGIN
Europe, Caucasus

CLIMATE AND HABITAT
Cool climate; fertile,
well-drained soil

LONGEVITY
Up to 400 years

SPEED OF GROWTH
20 centimetres–1.5 metres/
8 inches–5 feet per year

MAXIMUM HEIGHT
35 metres/115 feet

The leaves of the ash tree fall to the
ground when they are still green.

*Aleppo pine cones open slowly over
years, a process that can be sped
up if they are exposed to fire.*

Pinus halepensis (Pinaceae)

ALEPPO PINE

Greek Taste

The Aleppo pine is found in most of the lands that hug the shores of the Mediterranean – from Spain eastwards to Turkey, then south to Syria, the Lebanon, Jordan, Israel and Palestine. It was well after it had been identified as a pine that 'Aleppo' was added to its name, from the city in Syria where it was first described in detail. Although it is usually found at low altitudes, the Aleppo pine is known to grow at 1,000 metres (3,280 feet) above sea level in Spain and at 1,700 metres (5,600 feet) in North Africa.

It is an adaptable tree, existing happily where it is left to grow naturally, with a trunk bare for about 6 metres (20 feet) before branching out, or encouraged to spread from much nearer the ground. Its bark is thick and deeply fissured at the base of the tree, the fissures thinning in the higher parts of the tree. The leaves take the form of long, spiky, yellow-green to dark-green needles. The cones are green in youth, changing to a glossy red-brown by the time they are two years old.

The commercial value of the Aleppo pine lies in the nuts that are produced when the cones ripen and open. These nuts exude a resinous sap that was prized by the ancient Egyptians, who paid vast sums of money to buy it from neighbouring countries. It was used, like the resin of the lacquer tree, in the process of mummification. The ancient Greeks, however, had another use for the resin, in the production of wine.

One of the problems for early wine-makers was how to preserve wine. The solution was found to be to use Aleppo pine resin to seal the necks of wine vessels. Not all ancient wine-drinkers were happy with the flavour the resin imparted to the wine, however. Lucius Columella, a well-known writer on Roman agriculture in the first century AD, was highly critical of resin-infused wine. Despite that, some 2,000 years later, white or rosé retsina is considered by many to be the national drink of modern Greece.

Tectona grandis (Lamiaceae)

TEAK

Wood of Iron

A mature teak tree may grow to 45 metres (148 feet) in height and 2 metres (6½ feet) in girth. Surprisingly to most of us, it is a member of the mint family, making it a relative of the rosemary bush, basil and oregano – its leaves are used in the making of *Pelakai gatti*, jackfruit dumplings, in Kerala, India. It blooms from June to August, with sweet-smelling flowers that attract the bees that pollinate the teak's drupe. This ripens from September to December, and contains one stony seed embedded in fleshy pulp. The leaves are rough in texture, and the underside of each is covered with hairs.

The teak's natural home is in the monsoon forests of southern and southeastern Asia. A third of the globally produced teak timber comes from vast areas in Myanmar where the teak is farmed. Demand for teak has resulted in rapidly diminishing natural populations of the tree. In the nineteenth century, plantations were established in an attempt to capitalize on the growing demand, but exploitation continued. Today, illegal logging still takes place, driven by the demand for wood from large, mature wild trees, which is considered to be of a higher quality than that of commercially grown trees.

Teak timber has long been valued for its strength. Even when untreated, it is tight-grained, oily and resistant to decay. Ships made of teak are not subject to attack from the dreaded shipworm, a mollusc that has long destroyed other wooden ships. A shipwreck found in Oman, dating from the Harappan period (about 3300–1900 BC), was discovered to be made mostly of teak. The Chinese call teak 'ironwood', and discovered that burying it in soil for several years enhances its properties. This treated teak was used to build virtually indestructible junks. Its strength has been proved many times: when a teak hull collides with a steel hull it is often the steel hull that suffers more damage.

The plight of the teak tree still causes concern. Teak from sustainable plantations is now certified by the Forest Stewardship Council (FSC). Care should be taken when visiting garden centres for teak chairs, tables, benches and sun-loungers to furnish the patio. Always look for furniture stamped with the FSC accreditation – the teak tree will appreciate your concern.

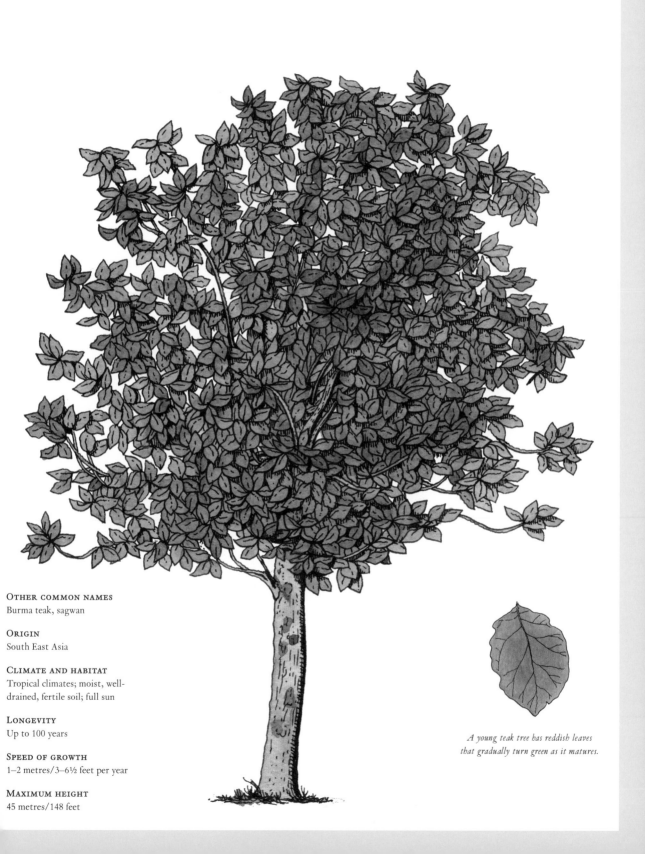

OTHER COMMON NAMES
Burma teak, sagwan

ORIGIN
South East Asia

CLIMATE AND HABITAT
Tropical climates; moist, well-
drained, fertile soil; full sun

LONGEVITY
Up to 100 years

SPEED OF GROWTH
1–2 metres/3–6½ feet per year

MAXIMUM HEIGHT
45 metres/148 feet

*A young teak tree has reddish leaves
that gradually turn green as it matures.*

OTHER COMMON NAME
Lebanon cedar

ORIGIN
Eastern Mediterranean basin

CLIMATE AND HABITAT
Typically north- and west-facing slopes 1,300–3,000 metres/4,300–9,800 feet above sea level, on free-draining alkaline soils. In some native ranges they can encounter extreme temperatures (-30°C to 30°C/-22°F to 86°F)

LONGEVITY
Up to 2,000 years

SPEED OF GROWTH
5–10 centimetres/
2–4 inches per year

MAXIMUM HEIGHT
40 metres/130 feet

The cedar of Lebanon produces its first cones when it is about 40 years old.

Cedrus libani (Pinaceae)

CEDAR OF LEBANON

Coveted by Pharaohs

The majestic cedar of Lebanon once thrived in the wilds of the Mediterranean basin, but thousands of years of demand for its timber in shipbuilding and construction decimated these ancient forests. Having no native trees of appropriate size, the ancient Egyptians coveted the cedars of their neighbours. Ancient Egyptian art reveals the importance of ships to the pharaohs and in Egyptian culture. Preparations for the afterlife went as far as burying entire vessels. Writing in the first century AD, Pliny the Elder describes the Egyptians as using the sap or oil of the cedar in the mummification process, a claim that has been verified by modern science.

There are many references to the cedar of Lebanon in the Old Testament. Psalm 92 alone has much to say about the tree: 'The righteous shall grow like a cedar in Lebanon… Put your trust in my shadow, and if not, let fire come out of the bramble and devour the cedars of Lebanon.' According to rabbinic literature, the prophet Isaiah, fearing King Manasseh, hid in a cedar tree, but the fringes of his garment gave him away. King Manasseh caused the tree to be sawn in half.

The cedar of Lebanon is a fine tree, a tall green conifer that grows rapidly until middle age (about 50 years), after which its growth slows steadily. It has a massive single trunk up to 2.5 metres (8 feet) in diameter. Its bark is rough and scaly, grey at first but turning black-brown, with deep fissures that peel in small chips. Its needle-like leaves are arranged in spirals, and it flowers in the autumn. It can take up to 40 years for cones to appear, and the young seed cones are resinous and pale green.

This is one of those special trees that play an essential identifying role in a landscape, like the palm trees of Miami, the oaks of Sherwood and the poplars that line roads in northern France. The cedar goes one step further, however: it is the national emblem of Lebanon, and is proudly displayed on the national flag. Historically, in the Ottoman Empire, which stretched from central Europe to the horn of Africa, the cedar was accepted in place of currency for tax payments.

Morus alba (Moraceae)

WHITE MULBERRY

Favourite of the Silkworm

The white mulberry started life in China, but spread across the world and is now cultivated in Mexico, Australia, North and South America, Turkey and the Middle East. The tree, which is deciduous in temperate regions but evergreen in the tropics, owes its popularity to the silkworms that feast in their thousands on its leaves. It is a handsome, fast-growing tree with dark brown bark, deep green leaves, white flowers and fruit that ripens through white, pink, red and black. A mature tree can grow as high as 18 metres (59 feet) but, by tree standards, the white mulberry does not have a long life. Most live about as long as a healthy (and lucky) human.

The fruit of the white mulberry is regarded by some as a superfood, high in fibre, protein and other nutrients. The bark is used in traditional Asian medicine as an antibacterial agent against food poisoning, and tea made from the leaf is considered to act as an anti-stress agent.

But the white mulberry is most famous as the silk tree. Legend has it that Leizu, wife of the Yellow Emperor Huangdi, invented the process of weaving silk in 2696 BC. While she was enjoying tea beneath a mulberry tree, a cocoon fell into her cup. She watched, fascinated, as a strong fibre unwound from the cocoon. Leizu saw its potential in making cloth, discovered how to combine the fibres into thread and, it is said, invented the loom. In reality, archaeologists have discovered remnants of silk in 8,500-year-old tombs at Jiahu, near the Yellow River. Each cocoon of the white mulberry caterpillar can yield as much as 300 metres (980 feet) of a thread that, woven into bolts of cloth, has played a major part in the culture and economy of China for thousands of years.

The tree has one exceptionally notable feature. When the time comes for it to release its pollen, the stamens that produce the pollen act as catapults, exploding and shooting the pollen out at 560 kph (350 mph) – about half the speed of sound.

OTHER COMMON NAME
Silkworm mulberry

ORIGIN
Central and East Asia

CLIMATE AND HABITAT
Ranging from cool, temperate
grassland through warm to tropical
forest, both dry and moist climates

LONGEVITY
80–100 years

SPEED OF GROWTH
10–30 centimetres/
4–12 inches per year

MAXIMUM HEIGHT
18 metres/59 feet

*Mulberry fruit start off white, before
ripening to pink, red and finally black.*

The leaves of the Ceylon ebony can be used as a blistering plaster.

OTHER COMMON NAME
Indian ebony

ORIGIN
India, Sri Lanka, Indonesia

CLIMATE AND HABITAT
Tropical, humid, lowland evergreen forest, often as an understorey species; acid to moderately alkaline soils of variable fertility

LONGEVITY
Up to 200 years

SPEED OF GROWTH
15–30 centimetres/
6–12 inches per year

MAXIMUM HEIGHT
25 metres/82 feet

Diospyros ebenum (Ebenaceae)

CEYLON EBONY

Cabinetmaker's Choice

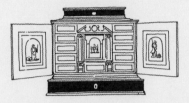

Like many slow-growing trees, the Ceylon ebony produces a hard wood. Indeed, ebony is twice as hard as oak. The tree is found mainly in Sri Lanka (previously known as Ceylon, hence the tree's name), Indonesia and southern India. It grows as tall as 25 metres (82 feet), based on a single thick trunk that branches when 8 metres (26 feet) in height. The sapwood is a light yellowish-grey, but inside that, the wood core is the glossy black associated with ebony. Where it grows in temperate climates, the Ceylon ebony produces a small fruit, similar to that of its relative the Japanese persimmon or kaki, *Diospyros kaki*. Its wood has almost no pitting and a naturally smooth, glossy surface, and is resistant to water and termites.

The ancient Egyptians valued ebony (which they called *hbny*), and there is evidence of trade taking place several thousand years ago. But the popularity of wood from the Ceylon ebony was at its peak from the sixteenth to the nineteenth centuries. Few trees could provide the world with the raw material for such a variety of uses. When polished, ebony becomes perfectly smooth, with a cold, metallic feel, and is hard enough to be finely carved and worked, making it ideal for decorative cabinets, chairs, boxes, desks and even tools and utensils. Among many other uses, ebony has been made into knitting needles, door and window handles, hooks, tripods, the black keys on high-class pianos (although the phrase 'tickling the ebonies' never caught on), chess pieces and chopsticks.

Ceylon ebony is still considered the best of all ebony woods for furniture-making. But the wood has become so valuable that it is now sold by the kilogram, and so popular that the tree's survival was threatened. It is now illegal in both India and Sri Lanka to sell ebony on the international market.

Commiphora myrrha (Burseraceae)

MYRRH

Worth its Weight in Gold

Myrrh is probably best known as one of the gifts brought by the Three Wise Men to give to the infant Jesus. It was most probably given in the form of fragrant resin, tapped from the myrrh tree, although the Gospel of Matthew – the only one of the four that identifies myrrh as one of the gifts – is not specific on that point. But, at that time, myrrh had been known for thousands of years as a perfume, an incense and a medicine that could be ingested when mixed with wine, or as a rub to reduce swelling and pain. It was also used in embalming, and as a means of masking unpleasant odours.

Adapted to the arid and exposed environment in which it grows – the Arabian Peninsula and parts of Africa – the myrrh is a tough, spiky tree, at its best in autumn, when the leaves change from green to yellow, then to pinkish-orange, and finally to red. It has a twisted trunk that is sometimes short and thick, at other times thin and lanky, with many branches, and above the trunk there is often a large umbrella-like canopy.

To release the resin, the bark of the tree is repeatedly cut deep into the sapwood, from which the resin bleeds with the consistency of gum. It hardens in three months, ageing from a clear, opaque pale yellow to a rich dark yellow, streaked with white. The pieces of myrrh resin are used raw in Chinese medicine.

Sahure, second pharaoh of the Fifth Dynasty from 2487 to 2475 BC, certainly appreciated that myrrh was indeed worth its weight in gold. In his last year as pharaoh he sent an expedition to the fabled land of Punt, which is believed to have been to the southwest of present-day Egypt. It was a trading nation, exporting gold, blackwood, ivory, wild animals and myrrh. The expedition returned with 80,000 measures of myrrh from which, it is said, Sahure established the first ancient Egyptian navy.

*The contorted branches of the myrrh
tree are covered in sharp thorns.*

Cherry trees are hermaphrodite – the female and male parts are found in the same flower.

OTHER COMMON NAMES
Sweet cherry, gean

ORIGIN
Northwestern and Central Europe

CLIMATE AND HABITAT
Deciduous woodland and hedges on fertile soil in mild to cold climates

LONGEVITY
70–100 years

SPEED OF GROWTH
30–60 centimetres/
12–24 inches per year

MAXIMUM HEIGHT
30 metres/98 feet

Prunus avium (Rosaceae)

WILD CHERRY

Red Veneer

Along with the sour cherry, *Prunus cerasus*, the wild or sweet cherry is an ancestor of all cultivated species of this fruit. To the poet A. E. Housman, the cherry was an inspiration. The second poem in *A Shropshire Lad* is almost a hymn of praise to this beautiful tree: 'Loveliest of trees, the cherry tree now/ Is hung with bloom along the bough…' The great Russian playwright Anton Chekhov gave the cherry tree the last dramatic moment in his play *The Cherry Orchard*. The curtain falls on the sound of axes felling the orchard, depicting the stupidity of both the old Russian aristocracy and the new Russian bourgeoisie.

There is no doubting the beauty of the wild cherry throughout the year. In winter, its purple-grey bark has a metallic sheen with rough brown breathing pores (lenticils). Quite apart from its magnificent covering of white blossom, in April its bright brown buds release long-stalked oval leaves with toothed edges that taper to a point. In autumn, these leaves change to shades of brilliant gold, orange and scarlet. The leaves of the cherry tree are also remarkable for two extrafloral nectar-producing glands (nectaries), outside the flower and at the base of the leaf, which contain a sweet substance that attracts ants which, in return, defend the leaf against insects.

After so much promise, the fruit of the wild cherry may seem a touch disappointing – to humans. It is sweet and sharp, and somewhat thick-skinned. Birds, however, thrive on its juicy pulp, stripping the trees in late June. It is hard to say where Pliny the Elder would have placed the cherry on his list of fruits ranging from *dulcis* (sweet) to *acer* (sharp).

Among the earliest evidence of the use of cherry wood are the remains of wooden platform dwellings in a prehistoric lagoon at what is now Eastbourne in Sussex, England. The wood used to construct this early settlement was oak, with stakes of wild cherry, and has been dated to 2400 BC. Nearly five thousand years later hard, reddish-brown cherry wood is still used in wood turning and as a veneer, for the spokes and legs of furniture, to make musical instruments and tables, and for tobacco pipes.

Prunus dulcis (Rosaceae)

ALMOND

Favours and Fertility

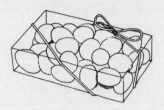

As befits a relative of the plum, peach and cherry, the almond tree produces drupes – stone fruits – that are rich in Vitamin B. In early spring pale pink blossom appears on its leafless branches. This becomes the firm, juiceless fruit, with flesh that is more leathery than fleshy, and which encases the shell of the almond.

Cultivated since 2000 BC, the almond has a long history. In the tomb of Tutankhamun, almonds lie among the thousands of other objects to provide for the king's afterlife. The Bible contains many references to the almond. It is mentioned in the Book of Numbers, Chapter 17, Verse 8: 'Moses went into the tent of the testimony, and behold, the rod of Aaron had sprouted… and it bore ripe almonds.' Ecclesiastes is less happy, however: there, the almond tree blossom is listed among the terrors that are to come in the evil days.

The almond tree is native to the Mediterranean regions of the Middle East, where it was one of the earliest domesticated fruit trees because of the ease with which it can be grown from seed. It is known to grow in far more temperate lands, but is cultivated mainly in the United States, Spain, Morocco and Australia. California is the world centre of cultivated almond trees. There, the pollination of the blossom is managed on a huge scale. Nearly a million beehives are trucked in from all over the United States to make sure there are enough bees to pollinate the almond blossom. As a result, some 2 million tons of almonds are harvested each year.

Almond nuts have many culinary uses: eaten whole, salted, toasted or blanched. Sugared almonds are a popular sweet in the Middle East, where they are known as *mblas* and represent the bitterness of life and the sweetness of love. Processed almonds can be made into almond butter and almond milk (to the delight of vegans), or mixed with honey and sugar to make nougat and marzipan. A variation of the species produces bitter almonds, which contain less oil, more water and a small amount of the poison hydrogen cyanide. Bitter almonds were once the traditional base of the Italian *amaretti* (almond macaroons).

OTHER COMMON NAME
Sweet almond

ORIGIN
South West and Central Asia

CLIMATE AND HABITAT
Mediterranean climate: warm, dry
summers and mild, wet winters on
cultivated land, scrub and rocky
slopes of moist, well-drained soil

LONGEVITY
60–80 years

SPEED OF GROWTH
40–80 centimetres/
16–32 inches per year

MAXIMUM HEIGHT
10 metres/33 feet

*A thick, juiceless green fruit –
known as a hull – encases the
shell of the almond nut.*

*Cinnamon-leaf oil is used in soaps,
creams and aromatherapy.*

OTHER COMMON NAMES
Ceylon cinnamon tree, sweet wood

ORIGIN
India, Sri Lanka

CLIMATE AND HABITAT
Tropical forests in moist,
well-drained soil from sea level
to 2,000 metres/6,500 feet

LONGEVITY
Up to 100 years

SPEED OF GROWTH
30 centimetres–1 metre/
12 inches–3 feet per year

MAXIMUM HEIGHT
20 metres/66 feet

Cinnamomum verum (Lauraceae)

CINNAMON

Ancient Goods

The popularity of cinnamon as a spice is recorded throughout history, largely because it was a precious commodity. The Egyptians were importing it as early as 2000 BC. Egypt stood at the crossroads of ancient spice routes and, as demand grew, Arab traders worked hard to maintain control of such a profitable spice; Pliny the Elder wrote of 350 grams of cinnamon being equal to over 5 kilograms of silver. The traders were successful, and for centuries the Western world had no idea where cinnamon came from. The Greek historian Herodotus (484–425 BC) believed that it was produced by the cinnamon bird; a thousand years later, the Crusaders were told it came from fish.

The origin of cinnamon spice is no longer a secret. It comes from the bark and leaf of the cinnamon tree. Commercially, in modern times, it comes from cinnamon trees that are coppiced by having their stems cut off at ground level. This promotes the growth the following year of a dozen or so new shoots. Over time the cinnamon tree begins to look like an upside-down wigwam. In the production of the spice, the inner bark of the tree is shaved into quills, which are dried and ground into powder. The leaves and twigs are steamed to produce cinnamon oil, although most oil and powder now comes from *Cinnamomum cassia*, the Chinese cinnamon.

The cinnamon tree is an evergreen, with oval leaves, a berry fruit and thick bark. In climates where winter temperatures drop no lower than a degree or two below freezing, the cinnamon grows as a handsome, ornate tree, with red-flushed young leaves that mature to a glossy mid-green.

The cinnamon has had an eventful history. A remorseful Emperor Nero, after murdering his wife in a fit of rage, ordered a year's supply of cinnamon to be burned. Worshippers of Bacchus enjoyed cinnamon as an ingredient in spiced wine. In the seventeenth century, having discovered a source along the coast of India, the Dutch bribed the local king to destroy it all, so that they could maintain their trade monopoly.

Hevea brasiliensis (Euphorbiaceae)

RUBBER TREE

Born in the Amazon

Left to itself, in the Amazon forests, the deciduous rubber tree grows up to 40 metres (130 feet), but it has rarely been left to itself because of its many uses as a unique material. It is the world's source of natural rubber, an essential component of modern life, from the soles of our shoes to the tyres of our cars, trucks, bikes and buses. As such, the tree is taken for granted, and many people believe that rubber is made in laboratories.

In order to tap the rubber tree to yield its latex (raw rubber), incisions are made deep enough to cause the tree to weep without damaging it. The tree produces the latex as a defence against attack by insects and herbivores. Through both its toxicity and its stickiness, latex can trap or immobilize the mouths of the insects. Tapping restricts and slows the cultivated rubber tree's growth, leading ultimately to a shorter life than that of the natural tree. As it ages, the tree produces less and less latex, and most cultivated trees are cut down when they are 25–30 years old.

Almost 4,000 years ago the Olmecs of Mesoamerica were using the sap of the rubber tree, to be fashioned into shoes or applied directly to the feet. Archaeological evidence reveals that it was also used to make balls of rubber for Mesoamerican games, which were as important to their culture as the gladiatorial contests were to the Roman Empire. Further evidence suggests that the ball games were every bit as brutal.

Europeans discovered latex much later. The first scientific paper on the subject was presented to the Académie Royale des Sciences in 1751 by the French explorer Charles Marie de La Condamine. In the nineteenth century entrepreneurs such as Charles Goodyear, Charles Macintosh and Thomas Hancock saw the commercial potential of rubber. Fortunes were made from the wild trees in the Amazon rainforest, meaning that few rubber trees were ever left alone.

OTHER COMMON NAMES
Para rubber tree, sharinga tree

ORIGIN
Amazon region (South America)

CLIMATE AND HABITAT
A middle-storey tree of tropical
rainforest in well-drained
soils, with access to water,
typically along riverbanks

LONGEVITY
Up to 100 years

SPEED OF GROWTH
6–15 centimetres/
2½–6 inches per year

MAXIMUM HEIGHT
40 metres/130 feet

*The rubber tree changes its
leaves during spells of dry
weather, usually twice a year.*

OTHER COMMON NAMES
Sandalwood, white
Indian sandalwood

ORIGIN
China, India, Indonesia, Philippines

CLIMATE AND HABITAT
Hemiparasitic; dry deciduous
forest; subtropical or tropical
climate; free-draining soils
excluding those of high alkalinity
and of shallow stony nature

LONGEVITY
Unknown (very few trees are left
undisturbed for more than 20 years)

SPEED OF GROWTH
10–25 centimetres/
4–10 inches per year

MAXIMUM HEIGHT
20 metres/66 feet

*Sandalwood flowers bloom all
year and attract ants and bees,
which pollinate the trees.*

Santalum album (Santalaceae)

Indian Sandalwood

Fragrant Smoke

The sandalwood tree is native to southern India, where its essential oil has been used for some 4,000 years. The ancient Egyptians used sandalwood in ritual burning and for embalming. One of the most important examples of sandalwood craft was found by archaeologists beneath the Great Bao'en Temple in Nanjing, China. An ornate sandalwood model of a stupa (a symbolic representation of Buddha), finished in silver and gold and embedded with jewels, was discovered in a stone chest. Inside the model there was a fragment of a human skull, believed by Chinese scientists to be that of Prince Siddhartha Gautama, the founder of Buddhism. It is also important in the Hindu Ayurveda, where the wood is used in the worship of the god Shiva. Sandalwood is also a wood essential to Jainism, Sufism, Zoroastrianism and to Korean, Japanese and Chinese religions.

The existence of the sandalwood tree in India is now threatened, though it now also grows in Pakistan, Nepal and, more recently, western Australia. The tree is the source of the second most valuable timber in the world, the most expensive being the African blackwood, once known as ebony. In Pakistan, the tree is owned by the government, and its use is strictly controlled not only for its wood, but also for its oil, recently priced at US$2,000 per kilogram. In 2009, Western Australia produced 20,000 kilograms of sandalwood.

For all its value, the sandalwood tree remains in danger. One barrier to the tree's conservation is its requirements as a hemiparasite. It augments its nutrient supply by feeding on neighbouring plants, especially such nitrogen-fixing species as *Acacia*. Another threat comes from its commercial value. The tree begins to develop its fragrant wood after only ten years of growth when, typically, it is uprooted and both trunk and roots are harvested.

The heartwood of the sandalwood is heavy, yellow and fine-grained, with a sweet, spicy and earthy aroma that persists for many years after the tree is felled. The small and appealing flowers appear from March to April and vary in colour from greenish-brown to red-brown or dark red. The bark is smooth and dark, almost black on young trees but matures to an ornate appearance, with deep vertical fissures.

Malus pumila (Rosaceae)

APPLE

Forbidden Fruit

It would take a good lawyer to put an end to the rumours (verbal and artistic) that the forbidden fruit of Genesis 3:5 was an apple. The list of suspects is long: grape, pomegranate, fig, carob, citron, pear and many others (including the mushroom). The ancient Jewish Book of Enoch suggests that the offending fruit came from the tamarind tree. One possible explanation for accusing the apple tree is that there has been a verbal misunderstanding, in which the word *malum*, meaning 'evil', has been confused with another Latin meaning for the word, 'apple'.

The apple is too good a tree to have produced the forbidden fruit. It is a generous tree. Its wood burns well in an open fire, and its pink-and-white spring blossom is not only beautiful in garden or orchard, but also the source of nectar for brood-rearing honey bees. Humans, animals, birds and insects love the flesh of its fruit. It is possible that it was the very first tree to be cultivated by humans.

Gardens of apples are mentioned in a Sumerian cuneiform manuscript dated around 1900 BC. Of a similar age is the archaeological evidence of rings of dried apples, possibly threaded on a string, found on saucers in Queen Puabi's grave at Ur, near present-day Nasiriyah in Iraq. The majority of botanists believe Kazakhstan's native species *Malus sieversii* is the source of the modern apple, of which there are now more than 3,000 varieties.

One protagonist of the apple tree has passed into American legend. John Chapman – known in folklore as 'Johnny Appleseed' – was a pioneer nurseryman who travelled from Canada through the northeastern states of America in the early nineteenth century, introducing the apple to local farmers and spreading the word (and the seed) of its beauty and benefits. The expression 'As American as apple pie' owes a lot to him.

OTHER COMMON NAMES
Paradise apple, and occasionally
older Latin names are used:
Malus domestica and *M. communis*

ORIGIN
Mountains of Central Asia

CLIMATE AND HABITAT
Open, mountainous conditions
and wooded hillsides; temperate
climate; a range of moist
and well-drained soils

LONGEVITY
At least 200 years

SPEED OF GROWTH
30–60 centimetres/
12–24 inches per year

MAXIMUM HEIGHT
9 metres/30 feet

*Most apple blossoms are pink
when they first emerge and fade to
white as the season progresses.*

OTHER COMMON NAMES
Italian cypress, Persian cypress

ORIGIN
Eastern Mediterranean,
North Africa, western Asia

CLIMATE AND HABITAT
Rocky soil, often with limestone
beneath, on slopes and in gorges,
occasionally in rock. Dry, hot
summers and winters with
varying amounts of rain

LONGEVITY
150 years; some examples
are believed to be more
than 600 years old

SPEED OF GROWTH
30–60 centimetres/
12–24 inches per year

MAXIMUM HEIGHT
20 metres/66 feet

*Small male and female cones
grow on the cypress, usually on
the tips of different branches.*

Cupressus sempervirens (Cupressaceae)

MEDITERRANEAN CYPRESS

Guardians of
the Underworld

The Mediterranean cypress has adorned its native landscape for thousands of years. It has traditional and mythical associations with the underworld, that dark place where souls go after death. This link probably results from the use of the tree's wood in antiquity. It was used for coffins in Egypt; the ancient Greeks kept the ashes of fallen soldiers in cypress urns; and some people believe that the cross used in the Crucifixion of Christ was made from cypress. In Mediterranean countries it is often grown in graveyards, where its longevity makes it a solemn, ever-present guardian.

The best-known archaeological evidence of cypress wood is that of the coffin of King Tutankhamun in about 1323 BC. At the time of his burial, fragments of the coffin lid were crudely chipped away to ensure a snug fit. Analysis of these chippings revealed them to be of cypress, testament to its durability. It had supported the immense weight of the sarcophagus for some 3,200 years.

The ancient Greeks and Romans appreciated this strength, using cypress for the doors of their palaces. It is believed that the original cypress doors of St Peter's Basilica in Rome lasted for a thousand years.

Often known as 'Fastigiata' or 'Stricta', to denote their pencil-like form, a number of selections of the cypress tree have become an iconic feature of Italian garden design, a fashion that began at the time of the Renaissance. Its narrow form makes it ideal for small gardens and narrow streets, and its remarkable tolerance of heat and drought has enabled it to remain undisturbed in the world's changing climate.

The Mediterranean cypress is another of those trees that are so stylish that they identify a landscape. On postcards, cinema and tablet screens, the narrow trunk with its dark green topping at once evokes the beauty and romance of Italy, Greece, Croatia, Turkey, Cyprus and a dozen more holiday havens. And Vincent Van Gogh immortalized the cypress of the village of Saint-Rémy in the foreground of his painting *The Starry Night* (1889).

One last plaudit for the cypress: its oil, distilled from the tree's young foliage and stems, is increasingly popular, and its light, spicy, refreshing pine scent is used in skin lotions and homeopathic treatments.

Ficus sycomorus (Moraceae)

SYCAMORE FIG

Life-giving

In ancient Egypt the sycamore fig was seen as sacred, a life-giving tree, its presence an indication of fresh water in apparently barren places, its fruit-laden branches providing shade and sustenance. Such bounty was considered a blessing from some benevolent power, typically the goddess Hathor, mother and consort of the sky god Horus and the sun god Ra. Both Hathor and the sycamore fig represented life and death, and were thus considered present in everyday life as well as in funerary rites and the journey to the afterlife. Indeed, the Egyptian king Tutankhamun was well supplied with sycamore figs after his death. They were found in the finely woven baskets that held many of the treasures that filled his tomb.

This was perhaps a sign of the honour in which Tutankhamun was held. Living in such arid lands, his people were always short of wood and took care in their use of timber from the sycamore fig. The ancient Egyptians used it in construction, in the manufacture of household and farming tools, and in the making of coffins. Today the wood is still used locally by artisans to make various implements, but less in construction, as it is vulnerable to termites.

The trunk of the tree is squat and thick. Above it is a vast pyramid of branches and leaves that mercifully create an area of shade from the sun. Its figs cannot compete commercially with fruit from the common fig tree (see page 18), but they are fleshy and sweet, and are enjoyed raw, stewed, dried and as an ingredient in alcoholic drinks. The orange-red figs grow throughout the year, supporting the life cycle of a symbiotic wasp on which the tree relies for pollination. The figs grow in dense clusters, tight against the main trunk and mature branches, where they present a curious but highly ornamental spectacle.

The sycamore appears in the Bible as 'sycomore'. According to Psalms it was almost completely destroyed in one of the Seven Plagues of Egypt; later the tax-collector Zacchaeus is said to have climbed one in Jericho to see Jesus passing on his way to Jerusalem (Luke 19:1–4).

OTHER COMMON NAMES
Mulberry fig, Pharaoh's fig

ORIGIN
North and East Africa,
South West Asia

CLIMATE AND HABITAT
A common savannah tree, often
found beside streams, rivers,
swamps and watering holes, or
singly in farmland; also inhabits
perimeters and clearings of
mountainous forests and evergreen
bushland; prefers deep well-
drained, fertile loam or clay soil but
will succeed in low-fertility soils

*The sycamore fig is slightly sweeter and
more aromatic than the common fig.*

LONGEVITY
Up to 700 years

SPEED OF GROWTH
1–1.5 metres/3–5 feet per year

MAXIMUM HEIGHT
20 metres/66 feet

*The versatile evergreen bay leaf can
be eaten fresh or dried, the latter
having a mildly peppery flavour.*

LONGEVITY
At least 60 years

SPEED OF GROWTH
20–40 centimetres/
8–16 inches per year

MAXIMUM HEIGHT
18 metres/59 feet

Laurus nobilis (Lauraceae)

BAY

Symbol of Victory

The bay was a tree of great symbolism to the classical Mediterranean civilizations, thanks to the largely vilified Philistines, a tribe best (or worst) known for their conflict with the Israelites, as documented in the Old Testament. Throughout history, the Philistines have been described as barbarians, but today many people, among them archaeologists from Bar-Ilan University in Israel, believe that they were responsible for an agricultural revolution in the Middle East, between 1200 and 600 BC. They are now credited with the introduction of many useful plants to the region, and with discovering the full benefits of such indigenous plants as the bay. The tree's use for food is likely to have been discovered by them.

In ancient Greece, bay is the classical laurel, sacred to Apollo as a symbol of glory, victory and honour. As such it was fashioned into wreaths to be worn by victorious warriors, the winners of sporting events and even great poets (hence the title 'Poet Laureate'). In addition, priestesses of ancient Greece ate bay leaves, which are known to be mildly narcotic, for the trance-like effect that allowed them to prophesy the future.

The Latin for 'laurel berry' is *baccalaureus*, and that word came to represent university undergraduates, who wore laurel-berry sprays and became known as 'bachelors', defined as unmarried men. The Romans also crowned their great men with laurel wreaths. Such was the respect for the laurel that its image was carved into Roman architecture in the belief that it would provide protection against evil spells, disease and lightning.

Today the bay leaf is widely used as a culinary herb. The bay itself grows happily in mild, temperate climates, in the garden, in tubs and containers, or as an indoor pot plant. Its oil is valued in massage therapy, reputed to alleviate arthritis and rheumatism; in aromatherapy it is used to treat earache and high blood pressure; and a poultice made of boiled bay leaves is a folk remedy for rashes caused by poison ivy and stinging nettles.

Theobroma cacao (Malvaceae)

CACAO

Bittersweet Commodity

Cacao is widely known as the source of chocolate. However, the sweet flesh of the cacao pod was first enjoyed as an alcoholic drink. A bottle dated as being in use in about 1100 BC, found among pottery on a site in Puerto Escondido in Honduras, contained cacao residue. Some 1,000 years later the delights of the cacao bean were discovered. Mayan hieroglyphs reveal the preparation of *chocolatl* or *xocolatyl*, a bitter, spicy drink taken for health and vitality. It was made by adding ground cacao beans, chillies, cornmeal and sometimes honey to the fermented flesh of the cacao fruit.

Art from this time depicts cacao as playing a major part in life for the wealthy, where it was used in dowries. Early Mayan records indicate that a woman had to prepare cocoa to prove that she could produce the proper froth. Every April, the Maya held a festival to honour the cacao god Ek Chuah, one of whose physical characteristics was a red-brown mouth, as though he was perpetually munching chocolate. The Aztecs greatly valued the cacao bean too, using it as currency – 80 to 100 beans could buy a new cloth mantle.

In 1502 Christopher Columbus introduced cacao beans to Spain. A few years later Spanish conquistadors arrived in Mexico, planning a monopoly of the precious bean. By 1634 the Dutch had managed to smuggle viable seeds out of Spanish territory, taking them to their plantations in Sri Lanka.

The invention of the cacao press in 1828 by the Dutchman Casparus van Houten Sr allowed the butter or fat to be separated from the roasted beans to produce cacao powder, which could then be remixed with the butter and with sugar to create a solid. The first chocolate bar as we know it today was made in 1847 at the Fry's chocolate factory in England, and in 1876 milk powder was added by the Swiss chocolatier Daniel Peter to produce milk chocolate.

Large cacao pods are of ornamental value. The clusters of small pink-white flowers are pollinated by a single species of tiny fly. As the fruit ripens to yellow-orange, their weight is borne by the main trunk and mature branches of the tree. So, the next time you open a box of chocolates, give a little thanks to the tiny fly that has made it all possible.

Other common names
Cacoa, cocoa

Origin
Mexico, Central America and
northern South America

Climate and habitat
Rainforest, underneath larger
evergreen trees; fertile, moist,
well-drained soils that are
moderately acid to alkaline

Longevity
Up to 200 years

Speed of growth
50–100 centimetres/
20–40 inches per year

Maximum height
8 metres/26 feet

*Large cacao pods encase the seeds
from which chocolate is made.*

The sweet chestnut fruit is encased
in a very spiny cupule called a burr.
The burr contains between one and
seven nuts, depending on the species.

OTHER COMMON NAMES
Spanish chestnut, Portuguese
chestnut, marron

ORIGIN
Southern Europe, Anatolia

CLIMATE AND HABITAT
Warm temperate climate;
deep, lime-free soil

LONGEVITY
Up to 2,000 years

SPEED OF GROWTH
10–25 centimetres/
4–10 inches per year

MAXIMUM HEIGHT
35 metres/115 feet

Castanea sativa (Fagaceae)

SWEET CHESTNUT

Straight is the Gate

The sweet chestnut probably existed before, and survived, the Ice Age, and was spread worldwide from a remnant population living in the Caucasus, its native home. The tree's modern botanical species name, *sativa* ('cultivated'), itself indicates that its distribution was largely human-assisted. Some of the earliest written references to it are from ancient Greece. It was most probably introduced from there to Rome and the Roman Empire, and in about AD 100 it reached Britain, where its nutritious nuts were mixed with a flour made from polenta to form a staple diet of the legionary soldiers.

It is a large, majestic tree, and belongs to the same family as oak and beech – just to remind us that it does, some species of oak and beech have similar leaves. The sweet chestnut can live for a long time. In many parts of Europe there are examples that are more than a thousand years old, gnarled, cracked and bulbous at the base to the extent that they look as though they have had almost every possible tree disease but have somehow survived. The Tortworth Chestnut, named after the small Gloucestershire village in England in which it lives, is thought to have been planted in AD 800, and there is written evidence of its existence dating back to 1150.

Unusually for nuts, sweet chestnuts are low in fat and protein, and in the Middle Ages – before the arrival of the potato from the New World – they were an important source of carbohydrate. The tree was also valued for its timber. The many woodlands containing sweet chestnut trees are testament to a past tradition of coppicing. The tree grows fast when young, producing long, straight poles. Its wood is also easy to split, and is still used to make hurdles, gates, fence poles and cleft palings. It is excellent firewood, but is inclined to spit, so be wary of it on open fires!

For centuries roasted sweet chestnuts have been sold on the streets at Christmas and New Year. Chestnut-sellers can still be seen in many English towns, hawking nuts roasted over braziers in autumn and winter.

Quercus ilex (Fagaceae)

HOLM OAK

Tree for Truffières

The foliage of the evergreen holm oak resembles that of the holly, hence the tree's name, from the old English word for holly: 'holm'. In its size, however, the majestic holm oak is more like the common or English oak (see page 136), living to a similar age and producing a similar timber. Although its natural home is in lands that have a Mediterranean climate, the holm oak grows happily in California and the southeastern United States. In southern England it resists the salt-laden winds that come off the sea, and an alien naturalized forest has grown up on the Downs overlooking Ventnor on the Isle of Wight. It was introduced there by the Victorians, whose passion was to collect and introduce new species to adorn their estates. With the aid of the jay – nature's feathered 'tree planter' – it spread quickly. The holm oak has so enjoyed its time at the seaside that it has seeded itself on wasteland, and goats have had to be employed to control its spread.

The tough timber of the holm oak was used by the Romans to supply spokes for their cartwheels and barrels for their wine. Earlier, the ancient Greeks used it to make crowns to be worn by the famous. They believed that the tree had the ability to tell the future, that its acorn represented fertility and that wearing acorn jewellery would increase fertility.

In more modern times, the holm oak was forested for the fungus that grows symbiotically on its roots. *Truffières*, forests of holm oaks, were planted in southern France to meet the growing gastronomic demand of Parisians for the black truffles, famously sniffed out from underneath the decomposing leaves by pigs or dogs. In Spain, the acorns, like those from the cork oak (see page 96), are an important food for pigs, giving Ibérico ham its unique flavour.

The holm oak retains its leaves for three years. The leaves have a thick, leathery texture that in winter allows them to store water that will be used to resist the summer drought.

OTHER COMMON NAMES
Holly oak, evergreen oak

ORIGIN
Southern Europe

CLIMATE AND HABITAT
A wide range of habitats; most
at home in milder maritime
climates and limestone soil;
extremely tolerant of salt

LONGEVITY
Up to 1,000 years

SPEED OF GROWTH
20–30 centimetres/
8–12 inches per year

MAXIMUM HEIGHT
25 metres/82 feet

*The acorns of the holm oak are
smaller than those of the English
oak, and have a pointed tip.*

OTHER COMMON NAME
Jewish lemon

ORIGIN
Eastern Himalayas (India)

CLIMATE AND HABITAT
Warm, wet climates; well-drained soils of limited fertility within a range of moderate acidity to alkalinity

LONGEVITY
Typically 50 years

SPEED OF GROWTH
20–40 centimetres/
8–16 inches per year

MAXIMUM HEIGHT
5 metres/16 feet

Citrons do not fall off the tree when ripe; they can grow up to 2.5 kilos (5½ pounds) if not picked.

Citrus medica (Rutaceae)

CITRON

Yuletide Flavour

Our enjoyment of the citrus fruits oranges, lemons and grapefruit goes back
so far that the origins of the species are shrouded in the mists of time.
Through dedicated archaeology and modern technology, it has now been
revealed that the small, shrubby citron tree began life in the foothills of the
eastern Himalayas in India, then travelled westwards. On the site of today's
Kibbutz Ramat Rachel in Jerusalem, archaeologists excavated the gardens of
a Persian palace dating from about 680 BC. Archaeobotanists analyzing plaster
removed from the garden's walls have found citron pollen dating from 538 BC,
an important period in Jewish history – the return of the Jews to Judah after
the fall of Babylon to the Persian king Cyrus the Great in 539 BC.

Citron is likely to have reached the ancient Greeks and Romans via
Persia, a country that was wrongly assumed to be its native source. The Greek
philosopher Theophrastus, sometimes referred to as the father of botany,
applied the name 'Persian apple' to citron in his *Historia Plantarum* (Enquiry
into Plants) of about 310 BC. Theophrastus dedicated the final chapters of
his book to the medicinal use of plants and trees, although he may have
exaggerated the power of the Persian apple as an antidote to poison.

A lesser-known citrus today, the citron's fruit is an oversized, wrinkly,
oddly shaped lemon. Weighing up to 2.5 kilos (5½ pounds), the ungainly fruit
is almost entirely rind and pith and, unlike its juicy relatives, produces only
a small amount of juice. The rind is valued for its fragrant oil, and, once the
bitterness has been boiled out, it absorbs sugar to form the candied peel that
is one of the ingredients in Christmas cake and pudding. The juice of the
citron is used in drinks or to make syrup for use in cooking, especially in Indian
cuisine. The Latin *medica* refers to the fruit's use in medicine throughout
antiquity, when its most common application was as a breath-freshener.

Having become a tree sacred to the Jewish people, a variety of citron with
smaller fruit, known as the etrog, is grown for use during Sukkot, the Feast of
Tabernacles, which occurs between late September and mid-October.

Prunus mume (Rosaceae)

CHINESE PLUM

Bringer of Spring

Amid the snow and ice of winter, the delicate, sweetly scented blossom of the Chinese plum appears as a harbinger of spring. The tree has several names. In the West it is known as the Japanese apricot or the flowering apricot, in Japan as *ume*, hence the delicacy *umeboshi* (pickled and dried plums). In China, where the tree originated, it is *mei*. Whatever its name, the tree has inspired the people of East Asia since the fifth century BC.

Its leafless branches blooming white or in shades of pink are depicted in paintings, fabric and ceramics, and its beauty has long inspired writers, poets and composers. In close competition with the much-loved peony, plum blossom is the official national flower of China, loved for its symbolic purity, its resilience and its perseverance in the face of adversity. Late in the Ming dynasty, the garden designer Ji Cheng wrote his definitive garden monograph *Yuanye*, in which he described the tree as 'the beautiful woman of the forest and moon'. In Confucianism, the plum blossom stands for the principles and values of virtue.

The tree's green or yellow oval or round fruit are small (no more than 3 centimetres – or a little over an inch – in diameter). Harvested in June and July, they are very sour but, after drying in the sun, they become food and drink in many forms. In China the fruit is popular in a thick, sweet sauce or, when smoked, boiled and sweetened, as a chilled summer drink. In Japan, the plums are pickled with salt, which brings out the fruit's sour and savoury flavours, and are also used to make a sweet wine, similar to cherry brandy.

The Chinese plum is cultivated in many forms: standard, weeping, ornamental and bonsai, in which form *mume* are encouraged to flower for the New Year, when the trees, studded with beautiful blossom, bring hope and the promise of joys to come. The Japanese love of the four seasons in nature is both spiritual and cultural, and the winter blossom of *mume* is the subject of the first flower-viewing of the year. This tree can be grown for its dainty flowers and sweet fragrance by gardeners in cooler temperate climates, where it delights as a sign of the end of winter.

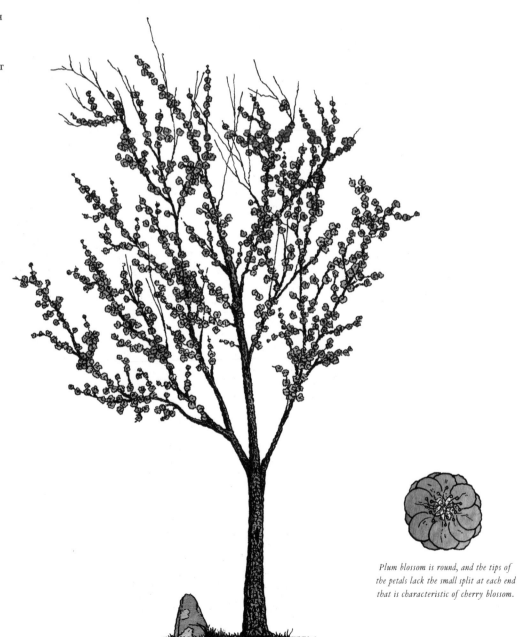

OTHER COMMON NAMES
Japanese apricot, meihua, mume

ORIGIN
China, Laos, Taiwan, Vietnam;
naturalized in Japan and Korea

CLIMATE AND HABITAT
Mountain slopes, sparse forests,
alongside streams and edges
of cultivated land; temperate
to subtropical climate; on
moisture retentive acid to
moderately alkaline loam or
clay soils that drain well

LONGEVITY
50–150 years

SPEED OF GROWTH
10–15 centimetres/
4–6 inches per year

MAXIMUM HEIGHT
10 metres/33 feet

*Plum blossom is round, and the tips of
the petals lack the small split at each end
that is characteristic of cherry blossom.*

OTHER COMMON NAMES
Indian banyan, banyan fig, banian

ORIGIN
India, Pakistan

CLIMATE AND HABITAT
Tropical forests; reliably moist,
acid to neutral, reasonably
fertile soil

*Large banyan leaves offer
welcome shade in the villages of
the Indian subcontinent.*

LONGEVITY
At least 700 years

SPEED OF GROWTH
Branches typically grow
20–40 centimetres/8–16
inches per year, while the
aerial roots are much faster

MAXIMUM HEIGHT
30 metres/98 feet

Ficus benghalensis (Moraceae)

BANYAN

Home to the Lac

Found throughout the Indian subcontinent's tropical forests, the curious yet incredible banyan tree has found a way of spreading that few other trees can match. The tiny seeds it produces are dispersed by birds. Those seeds that fall to the ground and germinate are unlikely to survive, but those that fall and germinate in the crevices or on the branches and stems of other trees – or even on human-made edifices – send roots down towards the ground. The roots take hold, and the young banyan slowly envelops and kills its host. Over time, these aerial roots form several buttressed trunks supporting large, dense leaf canopies; a single tree can grow to appear as a small forest. This process has led to the tree being known as the 'strangler fig', for a number of species of the genus *Ficus* have adopted this strange habit.

A native of the Ganges valley in northern India, the banyan takes its name from the Gujarati word *banya*, meaning 'grocer' or 'merchant'. To Hindus it is known as Kalpavriksha, the Tree of Life. It represents the Trimurti, the three lords of cosmic creation, preservation and destruction: Lord Vishnu is believed to be the bark, Lord Brahma the roots and Lord Shiva the branches. Although the Lord Buddha is said to have attained enlightenment while meditating beneath the Bodhi tree, *Ficus religiosa* (see page 90), he spent the following fifth week beneath the Goatherds' Banyan tree. Ground beneath the canopy of a banyan is often the site of prayer and meditation, and statues and temples may become completely covered by the tree. One such example of this process was discovered by students of Utkal University in the east Indian district of Khordha. Buried and hidden from view beneath a banyan, the students unearthed a rare 1,400-year-old statue of Lord Buddha with a protective seven-headed snake.

The banyan also plays host to *Laccifer lacca*, an insect that feeds on the tree's sap and secretes a resin to protect itself and its eggs. The resin is harvested and processed into shellac used as a coating for candles, in the production of nail varnish and in pharmaceuticals. The banyan has achieved recognition in the age of the computer, too. Its massive root system inspired the name for the computer network operating system known as 'Banyan Vines'.

Citrus sinensis (Rutaceae)

SWEET ORANGE

The Birth of the Orangery

From wild species of the southeastern Himalayan region, the sweet orange has developed through natural hybrids and human selection to become the world's most commonly cultivated fruit. The diversity of citrus makes it difficult to date the origins of the sweet orange, but the species is believed to be at least 2,500 years old. The oldest written reference to the orange is thought to be by the fourth-century BC Chinese poet Qu Yuan. His poem 'In Praise of the Orange' may well refer to the sweet orange or possibly to one of its parents, the mandarin orange. Cultivation of the sweet orange tree in the Mediterranean region came 2,000 years later, most probably brought there by Italian traders or Portuguese seafarers. During the sixteenth century, Spanish traders and explorers introduced the tree to the Americas, where production of the fruit became a vast industry.

Demand for the sweet orange grew so quickly throughout Europe that, in colder temperate countries, the rich built elaborate glasshouses, known as orangeries, in which to grow them. The first orangery was built in Padua, Italy, in 1545. The fashion spread. In 1617 an orangery was added to the Louvre Palace in Paris. This inspired Louis XIV to commission Europe's largest orangery at Versailles, where it reputedly contained 3,000 trees. In 1761 two orangeries were built in London, one at Kensington Palace and one in Kew Gardens.

The use of citrus fruit to combat the dreaded scurvy in sailors was discovered by the Scotsman James Lind, a physician and pioneer of naval hygiene. In 1753 Lind published his *Treatise of the Scurvy* with details of the many tests he had carried out on 12 sick sailors, using garlic, horseradish, mushrooms, cider, lemons and oranges. Those sailors receiving the citrus fruits were the first to recover.

The sweet orange does well under moderate temperatures. As well as being very beautiful, its white blossom produces a delightful scent and attracts many pollinating insects. Fruit and blossom can adorn the tree at the same time, and it has attractive deep green foliage all year round.

And finally… Nell Gwyn did indeed sell oranges in Restoration theatres, charging sixpence per orange, almost a day's pay to a soldier.

*Oranges are never found in the
wild – they are a hybrid between
a pomelo and a mandarin.*

OTHER COMMON NAMES
Sacred fig, peepal, bo

ORIGIN
India, South East Asia

CLIMATE AND HABITAT
Tropical rainforest to warm
temperate environments; adaptable,
preferring fertile alluvial soils

LONGEVITY
Up to 1,500 years

SPEED OF GROWTH
30–60 centimetres/
12–24 inches per year

MAXIMUM HEIGHT
30 metres/98 feet

*Like other members of the fig
family, the Bodhi tree needs a fig
wasp to pollinate its flowers.*

Ficus religiosa (Moraceae)

BODHI TREE

Enlightening Buddha

The Bodhi tree, or peepal as it is often known, is perhaps the most religiously sacred of all trees. It is native to India and parts of South East Asia, and is another member of the enormous fig family (see pages 18, 72 and 180). *Ficus religiosa* is a large wide-spreading tree, often deciduous in the dry season, occasionally semi-evergreen. It grows up to 30 metres (98 feet) tall, with a trunk up to 3 metres (almost 10 feet) in diameter.

Its remarkable leaves are heart-shaped, with a distinctive extended mouse-tail apex, or drip-tip, and a long, slender stalk that allows the leaves to move continuously even when there is no perceptible wind – a phenomenon that it shares with poplars and aspens, which have similarly shaped leaves. A large part of the root system of the Bodhi tree is above ground, forming a skeleton-like crinoline-shaped base to the tree. The fruits or figs of the Bodhi are small and green, ripening to red or purple.

The Bodhi tree is considered sacred by followers of Hinduism and Buddhism particularly. Hindus believe that the seemingly magic movement of the leaves signifies that Devas (heavenly, divine, terrestrial beings of high excellence) reside on them, although the prosaic explanation for the movement is the presence of hot air currents. In the Hindu scripture the Bhagavad Gita, literally translated as *Song of the Lord* and most often simply referred to as the Gita, Krishna says: 'I am the Peepal tree among the trees, Narada among the sages, Chitraaratha among the Gandharvas, and sage Kapila among the Siddhas.' Most famously, Lord Buddha attained enlightenment (*bodhi*) while meditating beneath the tree in present-day Bodhi Gaya in Bihar, India. The tree was destroyed and has been replaced several times, but a branch of the original was rooted in Anuradhapura, Sri Lanka, in 288 BC, and is known as Jaya Sri Maha Bodhi. It is the oldest dated angiosperm (flowering plant) in the world.

Ailanthus altissima (Simaroubaceae)

TREE OF HEAVEN
Poor Man's Silk

The common name 'tree of heaven' was originally applied to a tropical species of *Ailanthus* by the Ambonese people of Indonesia, derived from their word *ailanto* and denoting a tree tall enough to touch the sky. The name has been adopted by the West in relation to *A. altissima*, the only species hardy enough to grow in mild, temperate climates.

Few trees can compare with the tree of heaven for its good qualities, and fewer can compare with its bad qualities. On the bright side, it has ornate foliage, handsome bark and russet-coloured winged fruits. And it is tough. In her semi-autobiographical novel *A Tree Grows in Brooklyn* (1943), Betty Smith described it as growing 'in boarded-up lots and neglected rubbish heaps, out of cellar gratings', and as being 'the only tree that grows in cement'. In its native China, the tree of heaven has been used medicinally since at least the third century BC (it was listed among other trees in the *Erya*, the oldest surviving Chinese dictionary): its leaves used as a treatment for boils, abscesses and itches; and its bark to treat dysentery, internal haemorrhaging, epilepsy and even baldness. All this, and the tree is home to a species of silkworm that feeds on its leaves, producing silk that makes pongee cloth or Shantung silk, cheaper and more durable than that produced by the mulberry's silkworms.

But the tree of heaven has a dark side. In regions with long, hot summers, it aggressively expands its territory with a variety of weapons – setting seed, producing suckering growth and using a form of chemical warfare known as allelopathy. Toxins from its bark and leaves accumulate in the soil, inhibiting the growth of other plants. Eradication is difficult. The stumps of felled trees produce fast new growth, and it is capable of regrowing from fragments of root that remain in the soil.

In the West, the tree has been planted in parks and to line streets, perhaps unwisely. The Chinese name for the tree of heaven is *chouchun*, 'foul-smelling tree'. The odour of its male flowers has been graphically described as a mixture of burnt or rancid peanut butter, used gym socks and cat urine, hence, in many urban areas, its nicknames 'ghetto palm', 'stink tree' and 'tree of hell'.

OTHER COMMON NAMES
Chinese sumac, stinking sumac

ORIGIN
Northern China

CLIMATE AND HABITAT
High or low altitude; prefers moist,
free-draining, loamy soil; tolerates
all but waterlogged conditions;
mild to hot temperate climates

LONGEVITY
Up to 50 years (the suckers
of dead trees can live on)

SPEED OF GROWTH
50 centimetres–2.5 metres/
20 inches–8 feet per year

MAXIMUM HEIGHT
25 metres/82 feet

*The caterpillar of the
Ailanthus silkworm feeds on
the tree of heaven's leaves.*

OTHER COMMON NAMES
English holly, European
holly, common holly

ORIGIN
Europe, South West Asia

CLIMATE AND HABITAT
Warm to cold temperate areas;
prefers woodland shelter
on a range of soils that are
moist, yet well drained

LONGEVITY
Up to 500 years

SPEED OF GROWTH
10–25 centimetres/
4–10 inches per year

MAXIMUM HEIGHT
25 metres/82 feet

*In Britain, holly is used as a Christmas
decoration; in Germany, where it
is called* Stechpalme, *it stands in
for palm leaves on Palm Sunday.*

Ilex aquifolium (Aquifoliaceae)

HOLLY

Bush of Christ

Before holly was widely associated with the festival of Christmas, it was considered sacred and to have special powers by the Druids, who regarded it as a symbol of fertility and eternal life. Cutting down a holly tree would bring bad luck; conversely, hanging its dark green, leafy boughs in the home was believed to give protection. The Romans associated holly with Saturn, their god of agriculture and the harvest, and it was used decoratively during the December festival of Saturnalia. Today holly remains symbolic of Jesus Christ in two ways: the red berries represent the blood he shed on the day of his Crucifixion, and the pointed leaves symbolize the crown of thorns placed on his head before he died. In Germany, the holly is often known as *Christdorn*, 'Christ thorn'.

There are up to 800 species of holly, distributed across Europe, North Africa and western Asia. The common holly, *Ilex aquifolium*, is unquestionably the best known. More often than not, holly is seen in the form of an understorey evergreen near forest margins, as a hedgerow plant, or in its many cultivated forms as an ornamental shrub or small tree in gardens, parks and arboreta. It is especially comfortable in oak and beech woodland. Holly trees are dioecious (male and female flowers form on different trees, and only the female plants bear fruit). Holly berries, which are produced in midwinter, are an important part of the diet of small birds. Victorian gardeners greatly enjoyed finding comparatively rare strains of holly, such as those with golden, silver or variegated leaves.

Holly bark remains steel-grey and smooth throughout the life of the tree, no matter how big it grows. Both the sapwood and the heartwood are the whitest of all woods. Holly is a dense wood, long used to make bowls and chessmen, or for decorative inlaid work. When dyed black, it has been used as a substitute for ebony. Holly is also much used in traditional hedge-making, where its spines stop animals forcing their way through.

Quercus suber (Fagaceae)

CORK OAK

Preserver of Wine

In antiquity, sealing food and drink vessels to preserve their contents was a serious business, especially when the feeding of armies was involved. The most important and iconic of these storage vessels was the amphora, widely used by the ancient Egyptians, Greeks and Romans. Early attempts to seal amphorae used such materials as clay or leaves, secured with the resinous sap of pine and other trees, none of which could match the efficiency of cork. An early account of this use of cork comes from the second century BC, when the Roman senator and historian Cato the Elder strongly recommended sealing jars with cork and pitch after fermentation was complete.

Since then, the bark of the cork oak has long been familiar as the stopper in a bottle of wine, but cork has many other uses. It lies at the heart of every cricket ball, is an excellent fireproof insulator for heat, sound and vibration, can be made into footwear, is a hard-wearing material for flooring and gaskets, and is even used in the thermal protection systems of spacecraft.

Portugal and Spain produce well over half the world's harvest of cork. Portugal is the home of the 230-year-old Sobreiro Monumental Cork Oak, which stands in the town of Aguas de Moura and has been a Portuguese National Monument since 1988. The cork oak also plays an essential part in the Portuguese economy. From early May to late August, cork is removed from the outer bark of the tree without the use of machinery. With a specially shaped sharp axe, and cutting with great precision, skilled harvesters known as 'extractors' carefully remove this outer layer.

Climate change, outbreaks of disease and the increasing use of screw-top bottles for wine have all contributed to a decline in cork production, which is bad news for the extractors, and for the goats and pigs that feed off the acorns that drop from this remarkable, renewable tree.

OTHER COMMON NAME
Cork tree

ORIGIN
Mediterranean

CLIMATE AND HABITAT
Warm, temperate climate; free-draining, sandy soil; will tolerate periods of flooding but not excessive, prolonged freezing

LONGEVITY
100–300 years

SPEED OF GROWTH
60–90 centimetres/
24–35 inches per year

MAXIMUM HEIGHT
20 metres/66 feet

The leaves of the cork oak fall during the second year of growth; they are longer than those of the English oak.

Oil extracted from the flower of the neem tree is used in aromatherapy for its calming and relaxing effect.

OTHER COMMON NAMES
Nim, Indian lilac, Persian lilac, margosa

ORIGIN
Afghanistan, Pakistan, India, Sri Lanka, Bangladesh, Myanmar, China

CLIMATE AND HABITAT
Dry, deciduous forests; often germinates beneath and benefits from the protection of thorn bushes; temperate to subtropical climate; grows in a range of all but waterlogged soils

LONGEVITY
Up to 200 years

SPEED OF GROWTH
80–180 centimetres/
30–70 inches per year

MAXIMUM HEIGHT
30 metres/98 feet

Azadirachta indica (Meliaceae)

NEEM

The Healer

The neem, also known as nim or Indian lilac, is valued throughout its areas of natural distribution and beyond as a healing tree, and is a sacred tree in India. An evergreen, it has deep roots that help it remain vibrant despite the arid conditions in which it naturally occurs. It is used in Ayurveda, an ancient and popular form of medicine, whereby medicinal knowledge is shared by gods and human practitioners through sages. Among the earliest evidence of neem's applications is found in *Charaka Samhita*, a medical text in common use for 400 years from the second century BC.

When cultivated, the neem tree has a relatively narrow trunk, but in the wild, this thickens to support a magnificent spreading canopy made up of large sprays of bright green, narrow, serrated leaves. Delicate white flowers cover thin shoots in springtime, and these shoots hang and droop when the clusters of yellow-green fruits form. Mature trees, their bark distinctively fissured and plated in shades of grey, red and brown, can be seen lining many city streets to provide welcome shade.

For almost 2,000 years nearly every part of the neem tree has been used for medicinal or therapeutic purposes. One of the tree's many names in Sanskrit is *pinchumada*, destroyer of leprosy and healer of skin. In Ayurveda it is valued for its antibacterial, antiviral, antifungal and sedative properties. A traditional tincture of neem has potential for the prevention of malaria. Both Ayurvedic and Siddha practitioners use it for skin diseases and to detoxify the blood. A long-lasting use of neem is in oral hygiene, where one end of a twig is chewed to release its natural antiseptic. The chewed end divides into strands, making it an effective toothbrush.

Given such curative powers, it is not surprising that the neem tree has been worshipped by Hindu and Buddhist Indians alike. The deity Jagannatha, Lord of the Universe, is commonly represented as a temple icon carved from neem wood. The use of this wood and the form and features of these brightly coloured icons are full of meaning to the faithful, and the icons are reverently replaced every 12 to 19 years.

Ulmus minor 'Atinia' (Ulmaceae)

ENGLISH OR ATINIAN ELM

Supporting the Vines
of Rome

The origin of the English elm is now known to have been a single tree introduced by the Romans to England, and increased by means of genetically identical clones of the single tree, said to originate in Atina, Italy – hence the Atinian Elm. This unprecedented situation was made possible by the tree's ability to produce clones of itself in the root suckers, allowing it to regenerate naturally or by humans replanting them. Some 2,000 years ago the Romans transplanted the original clone from Italy for use in viticulture. Planting elms at regular intervals, they would cut back the tops at a height of 3 metres (10 feet), and the resulting twiggy growth provided excellent support for grapevines. The Atinian elm is now the tree's global name, although it is also still called the English elm.

Elm wood is strong, durable and water-resistant, qualities that made its timber useful in the manufacture of water pipes (from hollowed trunks), jetties, piers and lock gates on canals. Today, although it has a greater tendency than oak to shrink, it is prized for boats, flooring and furniture.

Coming from a single clone had disastrous consequences when a fungal disease, first identified by the Dutch, was spread by the elm bark beetle (hence known as Dutch elm disease). The disease is believed to have originated in Asia. It was first noticed in Europe in 1910, and subsequent waves, such as that in the 1960s, decimated elm populations. Today there are few sizeable specimens in Europe that have avoided infection. Two of the best are in a public park in Brighton, on the south coast of England. More than 400 years old, they are known as the Brighton Preston Twins, and every effort is made by the local authority that owns them to manage and protect these majestic specimens.

The elm was a favourite subject for the English artist John Constable, who included elm trees in two of his most famous paintings: *The Cornfield* (1821) and *Salisbury Cathedral from the Bishop's Garden* (1826). The tree also features frequently in English literature, most notably in Shakespeare's *A Midsummer Night's Dream*, where Titania addresses Bottom, saying: 'Sleep well and I will wind thee in my arms… the female ivy so enrings the barky fingers of the elm…'

Elm leaves are round to oval with toothed edges and a rough, hairy surface.

OTHER COMMON NAME
Field elm

ORIGIN
Southern and Eastern Europe
to North Africa, the Caucasus
and the Middle East

CLIMATE AND HABITAT
Banks of streams and rivers;
tolerates warmer, drier sites;
subtropical to temperate climate.
A pioneer species, tolerant of
waterlogging, salt, drought,
pollution and high winds

LONGEVITY
At least 400 years

SPEED OF GROWTH
15 centimetres–1 metre/6 inches–
3 feet per year (the higher speed
is for juvenile suckering growth)

MAXIMUM HEIGHT
30 metres/98 feet

OTHER COMMON NAME
Holy thorn

ORIGIN
Glastonbury (Somerset, England)

CLIMATE AND HABITAT
Water-retentive clay; full sun;
mild temperate climate

LONGEVITY
100–150 years

SPEED OF GROWTH
40–60 centimetres/
16–24 inches per year

MAXIMUM HEIGHT
7 metres/23 feet

*Every year at Christmas a
flowering sprig of the holy thorn
is sent to the British monarch.*

Crataegus monogyna 'Biflora' (Rosaceae)

GLASTONBURY THORN
The Miraculous Tree

Few trees have so dramatic and contested a history as the Glastonbury thorn, named after the town in Somerset, England, where the species is said to have originated. It is a form of the common hawthorn, *Crataegus monogyna*, which flowers in May. The Glastonbury thorn goes one further. It has a flush of growth in winter and produces a second flowering around Christmas (hence its name 'Biflora'). Viewed simply as a botanical curiosity, the thorn is a spiky and almost untidy-looking tree, with white flowers, and a trunk and branches that are thin and wiry. Nevertheless, the thorn has become the subject of myth, legend and religious faith over the last 2,000 years.

The first written record of this cultivar dates from the early sixteenth century, in an account of the life of Joseph of Arimathea, a wealthy Jewish man from Judea who, it is said, buried the body of the crucified Christ. According to legend, Saint Joseph subsequently travelled to Glastonbury carrying the Holy Grail of Arthurian folklore. On meeting the locals, who were unimpressed, he climbed Wearyall Hill, where it is said that he plunged into the ground a wooden staff that had belonged to Jesus. He then slept, and on awakening found that the staff had taken root and sprouted into a thorn tree, which became a national shrine and the beginnings of Christianity in Europe. The sacred tree went on to flower twice a year – once at Christmas and again at Easter – further enhancing its miraculous status.

With fame came abuse, however. Many cuttings were taken from the tree and its trunk was repeatedly carved and slashed. During the English Civil War, the alleged original Glastonbury thorn was destroyed as a relic of magic and superstition. Reshooting from the base, or being replaced by cuttings of the original, the tree survived into modern times, and it has become popular as a pagan symbol and the focus of New Age belief. Increasingly associated with witches and followers of Wicca (a Pagan new religious movement founded in 1954) and even, some locals fear, Satanism, the Glastonbury thorn was cut down by a chainsaw-wielding vandal in 2010. Happily, a sapling was planted nearby and the tree maintains its symbolic status and historical importance.

Prunus insititia (Rosaceae)

DAMSON

Croatian Favourite

Species in the genus *Prunus* provide a host of popular fruits – peaches, cherries, and almonds (see pages 28, 58 and 60) – but it is the 'plums' that display the greatest diversity. Despite more than 2,000 years of cultivation, the origins of many of these species, subspecies and cultivars remain mysterious. Even today, there is argument as to whether the origin of the damson was a cross between the sloe and the cherry plum, or whether the damson is a direct descendant of the sloe.

In warm climates the larger-fruiting cultivars in the genus are often dried and sold as prunes; in cool climates the fruit tend to be smaller and more suitable for cooking or preserving as jam. *Prunus insititia* falls into the latter category. It originated in South West Asia, but varieties are now found growing wild across Europe, and in North America, where they were introduced before the War of Independence.

The common name 'damson' is unique to Britain, as is the particular tree to which it applies. Damson stones have been discovered on Iron Age archaeological sites such as Maiden Castle in Dorset, a shock to those who believe the Romans introduced the damson tree to Britain. However, there is no doubt that the Romans were responsible for its cultivation and distribution. The damson trees of Britain are highly ornamental, with bright white flowers in early spring, followed by small dark blue and purple fruit with a dusky white bloom, but they are also tough enough to make good hedging and windbreaks.

An old rhyme tells of the slow growth of the damson: 'He who plants plums, Plants for his sons. He who plants damsons, Plants for his grandsons.' In 1575 the English writer Leonard Mascall praised the damson as the best type of plum, and advised 'gathering the fruit when ripe and drying them in the sun or above a hot bread oven in order to keep them for a long time.' It was the Victorians, and their access to sugar from the colonies, that made jam and fruit desserts not only possible, but also popular. In Slavic countries such as Croatia, damson varieties are used to make a local brandy known as *slivovitz*. An English nineteenth-century reference went so far as to say that the 'good damson wine is, perhaps, the nearest approach to good port that we have in England'.

OTHER COMMON NAME
Damascene

ORIGIN
South West Asia

CLIMATE AND HABITAT
Open woodlands, hedges and
the edge of cultivated land in
full sun or dappled shade; mild
temperate climate; grows in a
range of soils, particularly heavy
clay that is moderately alkaline

LONGEVITY
60–100 years

SPEED OF GROWTH
20–40 centimetres/
8–16 inches per year

MAXIMUM HEIGHT
6 metres/20 feet

*The raw damson plum is very astringent
and is best eaten once cooked with sugar.*

OTHER COMMON NAME
Indian mango

ORIGIN
India, Myanmar, Bangladesh

CLIMATE AND HABITAT
Subtropical climate, on fertile soil

LONGEVITY
Up to 300 years

SPEED OF GROWTH
20–60 centimetres/
8–24 inches per year

MAXIMUM HEIGHT
30 metres/98 feet

*The mango has a recalcitrant seed,
which means that the seed cannot
survive drying or freezing.*

Mangifera indica (Anacardiaceae)

MANGO

Colonial Pickle

Modern trading and refrigeration have brought the delights of the mango tree from the tropics, where it grows happily, to the rest of the world. The mango is one of the most widely cultivated fruits in the tropics, and is seen by some as the king of the drupes – although it should be remembered that as well as the cashew (*Anacardium occidentale*) and the pistachio (see page 38), the mango is related to poison ivy (*Toxicodendron radicans*). The sap from the tree should be avoided, as it can cause severe irritation.

The tree is native to the Indian subcontinent, where it is the national fruit of India and Pakistan, and the national tree of Bangladesh. In India mangoes have been cultivated for more than 2,000 years, but it wasn't until the fifteenth century that Portuguese sailors and traders in spices introduced it to Africa and South America. When ripe, the fruit is almost obscenely sweet, but mangoes do not travel well and should be eaten quickly. The first imports to America, therefore, arrived as pickles. Such was the popularity of pickled mangoes in the eighteenth century that the word 'mango' became synonymous with the pickling process, and other fruits, such as sweet or bell peppers, were often known as mangoes. The phrase 'to mango' came to mean 'to pickle fruit'. The mango remains popular in pickles and chutneys, but is more sought after in its sweet form, such as jams, juices and ice creams.

Culturally, perhaps the mango's greatest claim to fame comes from the calypso song 'Underneath the Mango Tree' (1962), which featured in the James Bond film *Dr No*. Written by Monty Norman, it was apparently sung by Ursula Andress as she emerged from the sea, conch shells in hand and knife slung at her side. In reality, it was sung by Norman's wife, Diana Coupland.

With its broad, spreading crown wide enough to stand beneath, the mango is an important shade tree. Legend has it that the Buddha himself was presented with a mango grove, that he might repose in its grateful shade. Much like the apple, hundreds of varieties of mango have been developed and are cultivated, where climate permits, throughout the world.

Ceiba pentandra (Malvaceae)

KAPOK TREE

Sacred to the Mayans

The ceiba or kapok is one of the tree wonders of the world. The Aztec, Maya and other pre-Columbian Mesoamerican cultures considered it sacred – a symbol of the link between heaven, earth and the world that was believed to exist below. This was a giant tree upholding the world, with roots reaching down into the underworld. According to the folklore of Trinidad and Tobago, deep in a forest there exists a huge kapok tree known as the Castle of the Devil. In it lives Bazil, the Demon of Death, imprisoned by a humble carpenter who had carved seven rooms out of the tree, and tricked Bazil into entering it.

The kapok tree is native to Mexico, Central America, the north of South America, the Caribbean and tropical West Africa. It is one of the largest flowering trees in the world, often exceeding 50 metres (165 feet) in height. An extended buttress of flattened roots, reaching 10 metres (33 feet) or more up the main trunk and as much as 20 metres (66 feet) across the ground, supports this wondrous colossus, enabling it to grow in shallow soil. One of the most remarkable features of the tree are the conical thorns that protrude from the young branches and trunks. These distinctive protuberances are a regular feature of Mayan art, portrayed on pottery incense burners and cache vessels. Many funerary urns have effigies of ceiba thorns on their sides.

However, it is for the cotton-like fluff obtained from its seed pods that the kapok tree is best known. The material is found inside the long, green, tamarillo-shaped fruit and surrounds the seeds to aid their dispersal by the wind. Kapok is waterproof and much lighter than cotton; it is most often used as a stuffing material, although it has now been largely superseded by synthetic fibres. It is now mostly grown commercially in the rainforests of South East Asia, particularly in Java, hence Java cotton, another of kapok's names.

The kapok is the national emblem of Guatemala and Puerto Rico, and also of Equatorial Guinea, on whose coat of arms and flag it appears. In Sierra Leone the tree is a symbol of freedom for the slaves that fled there.

OTHER COMMON NAMES
Java cotton, ceiba

ORIGIN
Mexico, Central and South
America, West Africa

CLIMATE AND HABITAT
Tropical rainforest; moist,
well-drained, fertile, acid
to neutral loamy soil

LONGEVITY
Up to 500 years

SPEED OF GROWTH
50 centimetres–4 metres/
20 inches–15 feet per year

MAXIMUM HEIGHT
60 metres/200 feet

*Conical thorns protrude from the young
branches and trunks of the kapok tree.*

OTHER COMMON NAMES
Candleberry, Indian
walnut, kukui nut tree

ORIGIN
Tropical Asia

CLIMATE AND HABITAT
Very adaptable; although associated
most with a tropical climate, it
has acclimatized to some mild
temperate areas; typically growing
on moist, well-drained, slightly
acidic soils of varying fertility

LONGEVITY
Unrecorded

SPEED OF GROWTH
30 centimetres–1.5
metres/1–5 feet per year

MAXIMUM HEIGHT
25 metres/82 feet

*Candlenut fruits are mildly toxic
when raw, but are used in cooking
in Indonesia and Malaysia.*

Aleurites moluccana (Euphorbiaceae)

Candlenut Tree

Hawaiian Light

The candlenut tree is one of an elite group of plants in the euphorbia (spurge) family, which includes the rubber tree (see page 64), the castor oil plant (*Ricinus communis*), Chinese tallow (*Sapium sebiferum*) and poinsettia. This is the fifth largest family of flowering plants, with huge economic importance both in the past and present. The tree is a medium to large evergreen with distinctive olive-green to silver foliage, variable in shape but, broadly speaking, maple-like. It is probably native to the Indo-Malaya region of Asia, but no one is sure of the true extent of its original habitat. What is certain is that it has been distributed extremely effectively by man across the tropics, both north and south of the Equator.

Nowhere has this tree been taken to people's hearts more than in Hawaii, where it grows wild and is known as *kukui*. It was probably introduced there by the first Polynesian settlers. In 1959, when Hawaii became the fiftieth US state, the candlenut was made the new state tree, owing to its importance to the island's people.

The tree is known for its fruit, which look a little like kiwi fruit. They are effectively capsules, each containing two seeds or nuts of great use, but not to eat – at least not raw, for they are extremely poisonous unless cooked. They have been more useful for providing light. Strings of dried nuts set alight burned for up to 15 minutes, so were used as a measure of time. The oil extracted from the seed has wide uses: to protect cotton bolls from insect attack; as a laxative; to waterproof paper; to varnish the surfboards for which Hawaii is famous; in soap-making; and in the manufacture of paint.

One of the most innovative uses is by fishermen, who chew the nuts and spit the resulting oily paste into the sea. This breaks the surface tension of the water, reducing reflections so that they can better see the fish below. But perhaps the most important use is in making traditional *leis*. The white flowers, big leaves and seeds are strung together to make these bead necklaces that are traditionally presented to visitors when they arrive or leave the islands, as symbols of friendship.

Myristica fragrans (Myristicaceae)

NUTMEG

Spoils of War

Myristica fragrans gives us two popular spices. First and foremost is nutmeg, with its rich, nutty, sweet flavour and aroma. Less used and less popular is mace, which has similar but milder properties. Together with the clove, a spice of the tree *Syzygium aromaticum*, they are native to the Banda Islands of eastern Indonesia. Until the nineteenth century, these islands were the only source of these three highly valued spices, and they were known as the Spice Islands.

Prized in antiquity for their medicinal qualities, nutmeg and mace were traded throughout Asia. Indian Vedic texts recommended nutmeg as a treatment for bad breath, headache and fever. Arabian cultures considered it an aphrodisiac, and in Elizabethan England, nutmeg was seen as a cure for bubonic plague. As a result, in the seventeenth century the nutmeg was considered, weight for weight, more valuable than gold.

Brought to Europe by Arab traders, nutmeg and mace were widely adopted in both medicine and cuisine, despite their great cost. The traders successfully hid the true source of the spices until 1497, when the Portuguese explorer Vasco da Gama rounded the Cape of Good Hope and broke the Arab monopoly in spice-trading. Following victory in wars against Portugal, the Dutch East India Company gained control of the Spice Islands and a monopoly of the spice trade, which it violently defended. Not to be excluded, the British comparatively peacefully took over the small island of Rhun from the Dutch. This contributed to a series of wars between the British and the Dutch, lasting until 1674, when Rhun was returned to the Dutch in exchange for New Netherland, now known as the state of New York.

The nutmeg is a handsome evergreen tree, with yellow pear-shaped fruit that splits open when ripe, revealing a striking seed or dark purple-brown nut covered in an irregular net-like vivid red aril – an extra covering of certain seeds. This seed or nut is typically ground to make the spice nutmeg, while the aril is dried and becomes mace.

ORIGIN
Banda Islands (Maluku, Indonesia)

CLIMATE AND HABITAT
Humid, volcanic, lowland forest

LONGEVITY
Up to 100 years

SPEED OF GROWTH
20 centimetres–1 metre/
8 inches–3 feet per year

MAXIMUM HEIGHT
20 metres/66 feet

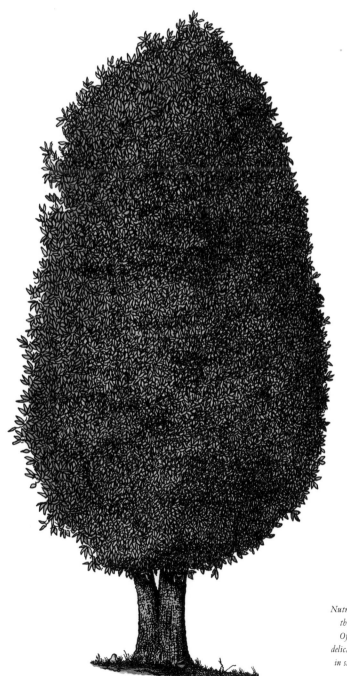

*Nutmeg and mace both come from
the* Myristica fragrans *nut.
Of the two, mace has a more
delicate flavour but both are used
in savoury dishes in East Asia.*

OTHER COMMON NAMES
Manitoba maple, elf maple,
ash-leaved maple, maple ash

ORIGIN
United States, Canada

CLIMATE AND HABITAT
Prefers a lowland habitat by
river or stream; frequently
colonizes wasteland or disturbed
ground; grows on a range of
moist soils that are neutral to
moderately acid to alkaline

LONGEVITY
Up to 100 years

SPEED OF GROWTH
15–60 centimetres/
6–24 inches per year

MAXIMUM HEIGHT
25 metres/82 feet

*One of the boxelder's names is ash-
leaved maple, because its compound
leaves resemble those of the ash.*

Acer negundo (Sapindaceae)

BOXELDER

Native American Music

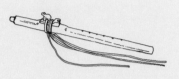

Acer negundo has many names, of which the most common is boxelder, although it isn't really an elder, but a maple. Its wood is similar in colour to box wood (see page 16), but nowhere near as durable. The foliage is similar to that of the elder or elderberry (*Sambucus* sp.). It is the only maple species with compound leaves that is native to North America, where in many areas it is considered a weed.

The harsh truth is that boxelder is a hard tree to love. It is fast-growing but short-lived. Its timber is low grade, and today mainly used as pulp for fibreboard. It is commonly infested with the boxelder bug (*Boisea trivittata*), a smart but smelly 1 cm (½ inch) black insect with a red edging. Clusters of these creatures often invade homes in autumn, when they seek a warm refuge for the winter.

However, back in the eighteenth century most Native American tribes made good use of the tree right across the continent, and not only as firewood. Some tribes used charcoal from boxelder wood for ceremonial painting and tattooing. Like that of the sugar maple (see page 144), the sap of the boxelder wood is rich in natural sugars, and most tribes learned to tap trees to make drinks and medicines as well as both syrup and crystalline sugar. Because the core wood of the young branches is pithy and soft, it was easily removed. The twigs could then be hollowed out to make pipes, bellows and flutes. *Acer negundo* has been identified as the material used to make Anasazi flutes, instruments that date from AD 620–70 and span a little over one and a half octaves, with a warm, rich tone. Examples of these ancestral Puebloan musical instruments were discovered intact during an archaeological dig in Arizona in 1931.

There are numerous ornamental selections of boxelder that have been selected and named over the years that now adorn parks and gardens. These include a subspecies from California in which the male form has showy pink flowers in pendulous tassels. There are various forms with variegated and golden foliage, too. A recently named selection called 'Winter Lightning' has bright yellow leaves and golden young stems, and can be coppiced for good effect in a winter garden.

LEMON

Bittersweet Thirst Quencher

Despite the fact that citrus fruit has been cultivated for 4,000 years, the origin and taxonomy of individual species and hybrids is still the subject of much debate. The ubiquitous lemon is a prime example, but recent genetic studies by Chinese researchers have revealed its parents to be citron (*Citrus medica*) and sour orange (*Citrus × aurantium*), both species from South East Asia. This latest knowledge provides evidence of where the lemon's journey into cultivation began.

As with so many trees, the Silk Road was instrumental in the lemon's distribution. Depictions of lemons in ancient Roman mosaics and frescoes were probably inspired by fruit traded with neighbouring Middle Eastern countries. The first written evidence of the lemon tree dates from the early tenth century, in an Arabic treatise on farming by Qustus al-Rumi, providing insight into agricultural innovations of that time. The early eleventh-century Persian poet, philosopher and traveller Nasir ibn Khusraw described Egyptians enjoying a citrus drink called *kashkab* that could be considered the origin of lemonade. It was made from fermented barley, mint, rue, black pepper and citron leaf. Later, the Jewish community of Cairo mass-produced and exported their popular drink *qatarmizat*, made from sugar and lemon juice. By the seventeenth century, lemonade (lemon juice, sparkling water and honey – a forerunner of the later *citron pressé*) had become so popular and fashionable in Paris that lemonade-makers formed a union, the Compagnie de Limonadiers.

Opening from red buds, the fragrant flowers of the lemon are white with reddish-purple tints on the back, and turn into the fruit so beloved by the patio grower in climates where growing lemons is more a matter of hope than of skill. But the citrus is relatively hardy. In cool climates it is possible to grow small lemon trees in ornamental pots if they are moved indoors in winter.

The last word on the fruit comes from Harry Belafonte's 1950s hit song 'Lemon Tree': 'Lemon tree very pretty, and the lemon flower is sweet, but the fruit of the poor lemon is impossible to eat.'

OTHER COMMON NAMES
None

ORIGIN
South East Asia

CLIMATE AND HABITAT
Tropical, subtropical and mild
temperate climates; moist, free-
draining, neutral to alkaline soil

LONGEVITY
Typically 50 years

SPEED OF GROWTH
10–60 centimetres/
4–24 inches per year

MAXIMUM HEIGHT
3 metres/10 feet

*The evergreen lemon produces fruit
all year round, and a single tree can
produce 270 kilos (600 pounds) a year.*

OTHER COMMON NAMES
None

ORIGIN
New Zealand

CLIMATE AND HABITAT
Subtropical and mild temperate climates in mixed forests subject to periodic volcanic disturbance; grows in a range of both wet and dry, reasonably fertile acidic to neutral soils

Tōtara seeds are succulent and red in autumn. Birds will eat them and then disperse the seeds as they void them.

LONGEVITY
Typically up to 1,000 years

SPEED OF GROWTH
5–25 centimetres/
2–10 inches per year

MAXIMUM HEIGHT
30 metres/98 feet

Podocarpus totara (Podocarpaceae)

TŌTARA

Heritage of the Maori

The tōtara is a canopy-topped conifer of the rainforests of both New Zealand's North and South Islands, although it is more common to the north. It is a majestic and ancient giant of prehistoric times. For more than 65 million years the tōtara's scaly, needle-like foliage has evolved into flattened, broad, greenish-brown leaves that are stiff and prickly to touch. This adaptation has been the key to the tōtara's success, allowing it to compete with the more highly evolved flowering plants that eventually dominated the landscape.

The Māoris, who first arrived in New Zealand from distant South Sea islands 800 years ago, prize the tōtara above all other native trees, chiefly for its timber, which is durable, resistant to rot and straight-grained, making it ideal for the making of spoons and other implements. It is a hardy species that will grow in almost any soil, and is the perfect building material for the Māori *waka* or canoe, which ranges from a few metres long, for fishing purposes, to the *waka taua*, a war canoe big enough to carry 100 warriors. Amazingly, whatever the size, most *wakas* are built from a single hollowed-out tōtara tree. Another use of the tree is *whakairo*, traditional carving which includes icons that are said to take on the character of those they represent. A fine early example of this art depicts Te Uenuku, the god of rainbows, carved between 1200 and 1500.

The tōtara's resistance to rot stems from its heartwood, which contains a chemical known as totarol. Its antimicrobial properties have been used by the Māori to treat fever, asthma and coughs. Although it is currently limited to use in cosmetics, its properties are subject to the attention of modern scientific research. Unfortunately, the arrival of European settlers, known by the Māori as Pākehā, resulted in most tōtara trees being felled for building or fence posts to enclose vast tracts of land for farming. The tree is now protected, and the use of its wood limited to that of dead trees. At least one tōtara is believed to be more than 1,800 years old. This is Pouakani, the oldest living tōtara. It has an impressively stout trunk, deeply fissured with the complex surface roots that grow to ensure the tree's stability during volcanic eruptions.

Sorbus torminalis (Rosaceae)

WILD SERVICE TREE
King's Crossbow

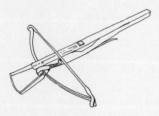

The wild service tree, despite being widely distributed across Europe, parts of Africa and northern Iran, is becoming scarce, threatened by changes in forest management and by the fungal disease fireblight. Once common in England and Wales – as a source of wood and for its fruit, which was a staple food of Neolithic man – it is now confined to ancient woodlands such as those of royal hunting forests. Early evidence of its existence in England is limited to that of archaeobotany, such as 8,000-year-old pollen found at Roundsea Woods in Cumbria, and charcoal recovered from Maiden Castle, an Iron Age hill fort in Dorset. Records of the service tree's historic value come much later in the form of documents in medieval Latin. An account from 1260 records that two service trees from Havering Park in Essex were taken to the Tower of London to make crossbows for Henry III. The wood, thanks to its fine grain, density and elasticity, has also been used in the manufacture of more peaceful objects, from musical instruments to the screws of a wine-press.

Another name for the tree in the United Kingdom is chequers tree, and its fruit are known as 'chequers'. The *torminalis* of its Latin name means 'useful against colic', and the small round or pear-shaped, reddish-brown, speckled fruits were used regularly to make or flavour alcohol, which links the tree to the many pubs in England and Wales called 'The Chequers'. The distinctive black-and-white chequerboard hangs over pubs and inns to this day, although in most cases the origin and meaning of the name have been lost. The fruit is sour at first, ripening to sweetness after 'bletting' by the frosts of autumn, when it becomes attractive to both humans and birds. The fruit was traditionally picked in bunches and hung above the hearth to ripen. Similar in taste to dried apricot, the soft, sweet fruit was enjoyed straight from the bunch or as a dessert ingredient.

The tree itself is good-looking, with profuse white blossom in May. In autumn it has a fiery display of rich copper-red leaves, similar in shape to those of the maple, and which are a popular food for moth caterpillars.

Other common name
Chequers tree

Origin
Most of Europe, eastern
Mediterranean, North
Africa, the Caucasus

Climate and habitat
Open oak or ash woods, hedgerows
on clay or lime soil in full sun;
tolerant of drought as well as
periods of waterlogging

Longevity
At least 100 years

Speed of growth
10–25 centimetres/
4–10 inches per year

Maximum height
25 metres/82 feet

*The wild service tree is hermaphrodite,
so both female and male reproductive
parts are contained within each flower.*

OTHER COMMON NAME
Cocoanut

ORIGIN
South East Asia, southern
India, Sri Lanka, the Maldives,
Lakshadweep (India)

CLIMATE AND HABITAT
Full sun on tropical and subtropical
seashores; occasionally extending
inland on alluvial plains

LONGEVITY
Up to 100 years

SPEED OF GROWTH
30–90 centimetres/
12–35 inches per year

MAXIMUM HEIGHT
30 metres/98 feet

*Coconuts can float extraordinary
distances across oceans before putting
down roots when they reach land.*

Cocos nucifera (Arecaceae)

COCONUT PALM

Tree of a Thousand Uses

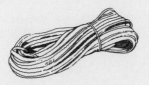

Throughout tropical regions the coconut palm is primarily a tree of commerce. To those who live in the rest of the world, however, it is a romantic icon of exotic beaches in faraway places, instantly associated with pristine white sand and warm turquoise seas.

The origin of the palm has always been a matter of debate, since its dispersal and subsequent cultivation have, over thousands of years, been aided by ocean currents carrying its floating seeds, as well as by humans. Recent DNA analysis has revealed two separate areas of early cultivation, the Pacific basin and the Indo-Atlantic Ocean basin, spreading east from South East Asia and west from southern India, Sri Lanka, the Maldives and the Lakshadweep Islands, off the southwestern coast of India.

Every part of the tree is used by humans. First and foremost, the coconut – botanically not a nut but a drupe – is the perfect portable source of water and food, while the shell can be burned as fuel and the fibre made into coir, which is used as rope, in upholstery and flooring as well as providing a potting medium for plants. The leaves and trunk are used in construction, while the sugary sap can be fermented into an alcoholic drink. The flowers are cooked as a vegetable, as is the leafy central bud, although removing it kills the tree. The pith of the trunk is made into bread or added to soup and other dishes, and the roots are roasted to make a coffee-like drink.

Little wonder, then, that the Malaysians call the coconut palm *pokok seribu guna*, 'the tree of a thousand uses'. Its value has also been appreciated by Europeans, and the Italian merchant and explorer Marco Polo first named it *Nux indica* (Indian nut) in 1280. The name 'coconut' is believed to have originated with Spanish and Portuguese explorers, who saw the 'nut' as 'El Coco', a mythical, hairy monster of their culture.

Today the coconut industry supplies a growing demand throughout the world, especially for food and cosmetics. The tree's adaptation to coastal situations, along with its tolerance to drought, means that this architectural palm is widely used to ornament streets and landscape gardens in warm, frost-free climates.

Coffea arabica (Rubiaceae)

COFFEE

The Stimulating Tree

Until some 600 years ago, *Coffea arabica* was merely an understorey tree of Ethiopia's mountain forests. Today it and other *Coffea* species are the world's most economically important trees, second only to crude oil in international, plant-based commerce.

The discovery of the world's favourite drink is subject to myth. The most popular story concerns Kaldi, an Ethiopian goat-herder. Kaldi observed that his goats were energized after they had eaten certain red berries. He tried the fruit himself and was similarly invigorated. A monk observing this reaction was curious to learn more and spent a sleepless night back at the monastery. Knowledge of the energizing berries spread.

There is evidence of coffee-trading in the Yemeni part of Arabia in the fifteenth century, where *coffea* beans were first boiled, producing a beverage they named *gahwa* – 'that which prevents sleep'. Cultivation on the Arabian peninsula and subsequent trade resulted in coffee becoming established in Persia, Egypt, Syria and Turkey. In 1475 the world's first coffee shop, Kiva Han, opened in Constantinople (Istanbul). Early coffee houses were named 'schools of the wise', since they became important hubs for the exchange of news, ideas and culture in general. The annual pilgrimage of Muslims from across the world to Mecca was instrumental in the early distribution of coffee. In 1511 Kha'ir Bey, governor of Mecca, unwisely attempted to ban coffee. The Sultan subsequently decreed that coffee was sacred and had the governor executed for embezzlement.

The coffee tree prefers to grow at higher altitudes in the frost-free moist climate of the tropics and subtropics where there is a short dry season and deep, well-drained soil. It crops best when between five and ten years old. Arabica is considered the highest-quality bean among *Coffea* species, and accounts for over half the world's production. Most of the rest is the robusta bean *Coffea canephora*. Although the high level of caffeine in the tree is believed to repel insect pests, a recent study suggests that pollinators visiting the white, sweetly scented flowers might become, like humans, hooked on caffeine. Perhaps that shows that the caffeine is an evolutionary trait on the part of the tree, to encourage return trips from these beneficial visitors.

OTHER COMMON NAMES
Arabian coffee, Arabica
coffee, Kona coffee

ORIGIN
Ethiopia, South Sudan

CLIMATE AND HABITAT
Humid evergreen forest in
a tropical climate, 600–700
metres/2,000–2,300 feet above
sea level; rich, slightly acidic soil

LONGEVITY
At least 100 years

SPEED OF GROWTH
10–30 centimetres/
4–12 inches per year

MAXIMUM HEIGHT
8 metres/26 feet

*The red fruit surrounding the
coffee bean is full of antioxidants,
and is increasingly being
used in health products.*

OTHER COMMON NAMES
Sweetgum, redgum, star-leaved gum, alligator wood

ORIGIN
United States

CLIMATE AND HABITAT
Warm temperate climate; low-lying; fertile, moist soil

LONGEVITY
Up to 400 years

SPEED OF GROWTH
15–60 centimetres/
6–24 inches per year

MAXIMUM HEIGHT
30 metres/98 feet

The leaves of the American sweetgum are maple-like, and turn to rich flame colours in the autumn.

Liquidambar styraciflua (Altingiaceae)

AMERICAN SWEETGUM
Liquid Amber

Liquidambar styraciflua – also known as redgum, star-leaved gum, alligator wood or sweetgum – is native to the eastern United States, where its distinctive star-shaped, dark green foliage turns in the fall from yellow to gold, deep orange, fiery red and finally deep purple before, reluctantly, the tree lets them drop, after all other species are already bare.

The bark is furrowed, rough and scaly, hence the name alligator wood. The tree, however, is best known for its seed pods, which are sometimes called gumballs. Indeed, the tree's genus name, *Liquidambar*, which was given to the tree in 1753 by the Swedish father of botanical taxonomy, Carl Linnaeus, derives from the Latin *liquidus* (fluid) and the Arabic *ambar* (amber), alluding to the fragrant gum the tree exudes. The species name, *styraciflua*, is an old name meaning 'flowing with storax', another reference to the gum.

The first mention of any use of the amber is by the Spanish conquistador Juan de Grijalva, who explored the coast of Mexico and conquered and ruled Cuba on behalf of Spain. De Grijalva writes of exchanging gifts with the Maya in 1517. The Maya, who used the gum for medicinal purposes, presented the Spanish explorers with hollow reeds filled with herbs and a sweet-smelling amber liquid that, when ignited and smoked, gave off a pleasant smell.

News of the American sweetgum's beauty and uses spread to Europe. In 1681 it was introduced to England by John Baptist Banister, a missionary chaplain and naturalist sent to North America by Henry Compton, then Bishop of London, whose diocese included the American colonies. The Bishop was a tree-lover, and he planted the sweetgum in the gardens of his palace in Fulham, southwest London.

After oak, the hardwood of the American sweetgum is the second most commercial wood in the United States, where it is used to make flooring and furniture, cabinets and panelling, as a veneer, and in the making of baskets, barrels, bowls and boxes.

Pinus sylvestris (Pinaceae)

SCOTS PINE

Caledonian Beauty

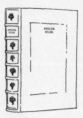

Pinus sylvestris is the most widespread species of pine. Of the 125 or so species of pine in the world, most are in the northern hemisphere, across Europe and northern Asia, but a few tiptoe over the Equator. The Scots pine is arguably the most useful, as well as one of the most beautiful. One of its loveliest distinguishing features is the shading of its bark, which is pinkish grey in the young tree and orange in the old; even the grey twigs show a tendency to turn russet.

It has long been valued in Scotland, where it forms the remnant of pine forests that covered much of the British Isles thousands of years ago. The tree spread after the Last Glacial Maximum, the point in the last Ice Age when the ice sheets were at their greatest. It came to England from France 9,000 years ago and arrived in Scotland a few hundred years later from a different place, probably Scandinavia or Ireland. As the climate warmed, it became extinct in most of the British Isles and now occurs as a natural remnant only in the glens and valleys of Scotland, and in the Caledonian Forest, a priority habitat under the UK Biodiversity Action Plan.

The Scots pine grows mostly on acidic and poor sandy soils, often with the silver birch (see page 204). The fauna of these mixed woods is uniquely rich with pine marten, wildcat and red squirrel, the capercaillie, crested tit and Scottish crossbill, the Scottish wood ant and Rannoch Looper moth – species unlikely to be found in forests to the south.

Timber from the Scots pine, a strong softwood, is widely used in the construction industry, and the manufacture of fencing, gate posts and telegraph poles. Once upon a time, it was used to make pit props for coal mines. The tree is also tapped for resin to make turpentine, and its inner bark can be made into rope. Dry cones make a fine kindling.

Mentioned in Shakespeare's *Richard II* and in *Sylva, or A Discourse of Forest-Trees and the Propagation of Timber in His Majesty's Dominions* (1662), the influencial text on forestry by famed writer and gardner John Evelyn, the forests of Scots pine were planted by wealthy landowners, while Jacobite supporters, it is said, planted their own Scots pines to show loyalty to their cause.

*As with most conifers, the Scots pine
is pollinated by wind. Cones take two
years to mature from the female flowers.*

OTHER COMMON NAME
European redbud

ORIGIN
Southern Europe, South West Asia

CLIMATE AND HABITAT
Temperate climate;
free-draining soil

LONGEVITY
Up to 300 years

SPEED OF GROWTH
10–20 centimetres/
4–8 inches per year

MAXIMUM HEIGHT
12 metres/40 feet

*Bright pink pea-flowers blossom
before the leaves between March
and May, and later deep purple
pods appear on the branches.*

Cercis siliquastrum (Fabaceae)

JUDAS TREE

Tree of Blood

There is little or no evidence that Judas hanged himself on a tree that now bears his name, or indeed any other tree if the Acts of the Apostles is to be believed. According to this book of the Bible, an unnamed man betrayed Jesus, bought a field with the 30 pieces of silver reward, and, 'falling headlong, he burst asunder in the midst, and all his bowels gushed out'.

But the legend and the tree at the centre of it remain. The Judas tree has a liking for a dry climate and is native to Spain, southern France, Italy, Bulgaria, Greece and Turkey. It is a small tree with a narrow trunk topped by a spreading crown. It is covered in early spring with a profusion of magenta-pink flowers, which appear before the leaves. This gem of a tree has been in cultivation in the British Isles for more than 300 years. Some of the oldest and best specimens are to be found in the Cambridge University Botanic Garden. Like most of its kin in the legume family, it produces its seed in pods. These are often brightly coloured, last well into the autumn and then persist on the leafless branches throughout the winter. The flowers of *Cercis siliquastrum* are edible, and can be eaten in mixed salads. They figure frequently in sixteenth- and seventeenth-century herbals, but are rarely used in this capacity today.

The name 'Judas tree' may derive from the French *arbre de Judée*, 'tree of Judea', a reference to the region in which the tree was once thought to grow wild. There is the myth that Judas Iscariot hanged himself from a tree of this species in the *Hakeldama* (the Field of Blood), in the Valley of Hinnom surrounding the old city of Jerusalem. Another origin of the name comes from the flowers appearing on bare branches, often directly on the trunk, making the tree apparently ooze blood – also symbolic of Judas' possible suicide.

There are many notable selections of this species. The unusual white-flowered form is especially beautiful, and gained an Award of Merit from the Royal Horticultural Society when it was exhibited in 1972. The variety 'Bodnant', named after the garden in North Wales where it was first observed, is particularly fine, with deep reddish-purple flowers and maroon seed pods.

Carya illinoinensis (Juglandaceae)

PECAN

Illinois Nuts

The pecan tree is deeply entwined with the history of the United States. It was held in high esteem by the nation's founding fathers. The great plantsman, landscape gardener and first United States president, George Washington, was given a bag of pecan nuts, which sprouted successfully at his garden at Mount Vernon, Virginia. The bag came from the man who was the primary author of the Declaration of Independence and later third president, Thomas Jefferson. In his orchard at Monticello, also in Virginia, Jefferson had several trees that both he and Washington referred to as 'Illinois Nuts'.

The pecan is a large deciduous hickory tree (see page 160), native to the southern United States and Mexico. It is so closely related to the walnut tree (see page 36) that Spanish settlers referred to pecans as *las nuez* (walnuts) when they introduced them to Europe, Africa and Asia in the sixteenth century. Prior to colonization by European settlers, pecans were important currency for bartering, as well as nutritious food for the indigenous population. The nuts kept well for at least a season, which was a bonus as the trees do not produce good crops of fruit annually.

The pecan is one of the more recent commercial crops, starting its important role in the American economy in the 1880s. Annual world production now exceeds 135,000 tons, split more or less equally between Mexico and the states of New Mexico, Georgia and Texas. Modern clonal selection has to a large extent overcome the natural biannual cropping by providing clone pairs that are compatible with each other in alternate seasons.

The pecan is the state tree of Texas (the town of San Saba claims to be the pecan capital), and several towns in the southern states host annual events to celebrate the pecan harvest each year. The long-living tree has other uses, too. It has an amazing grain that makes it much in demand for furniture-making and as a floor-covering, and it provides welcome shade in some of the hottest parts of North America. And, perhaps most of all, the pecan nut makes a wonderful filling for a pie.

OTHER COMMON NAME
Illinois nut

ORIGIN
United States, Mexico

CLIMATE AND HABITAT
Subtropical to warm
temperate climate; rich,
moist, free-draining soil

LONGEVITY
Up to 300 years

SPEED OF GROWTH
30–60 centimetres/
12–24 inches per year

MAXIMUM HEIGHT
40 metres/130 feet

*The pecan grows inside a green drupe,
or fruit. It is one of the world's
most nutritious nuts, containing
19 minerals and vitamins.*

*Plane tree seed pods hang in bristly
clusters and often remain on the
tree after the leaves have fallen.*

OTHER COMMON NAME
Old World sycamore

ORIGIN
Eurasia

CLIMATE AND HABITAT
Mild temperate to subtropical
climates; damp soil in low-
lying areas; adaptable to drier
situations once established

LONGEVITY
Typically 500 years

SPEED OF GROWTH
15–60 centimetres/
6–24 inches per year

MAXIMUM HEIGHT
50 metres/165 feet

Platanus orientalis (Platanaceae)

ORIENTAL PLANE
Shade for the Soldier

The plane trees (*Platanus*) are unique to their own family, and most of the few species are native to North America. They are all large trees, ranging from 30 to 50 metres (98 to 165 feet) in height. There are two principal species in the northern hemisphere. The most common North American species is *P. occidentalis*, confusingly known as the American sycamore. The other, *P. orientalis*, is native to Asia and Europe, from the Balkans at least as far east as Iran – one specimen was thought to be the Tree of Hippocrates, beneath which he taught medicine on the island of Kos. Although some species adapt to drier and more polluted environs, the trees are happiest in wetland or by rivers, where they are often seen in the company of alder, willow and poplar. The plane is very important in Persian gardens, which are designed around water and shade. In Greece it is a venerated shade tree of great age, many examples being over 500 years old.

The most common plane tree is a chance hybrid of *occidentalis* and *orientalis*, usually referred to as the London plane (*P.* × *acerifolia*), and is believed to have first been identified in Oxford in the early 1670s. It grows quickly, has the health and vigour of a good hybrid and is resistant to pests and diseases. In its eponymous home city it is one of the most commonly planted street trees, where it endures, soaks up and even appears to thrive on pollution. As the old dark plates of its bark fall from the trunk, dappled patches of yellow and olive-green appear.

Legend has it that in the late eighteenth century Napoleon Bonaparte ordered the planting of vast avenues of these trees to create shade for his marching armies. Whatever the truth, these avenues are now threatened with removal because of the large number of casualties suffered by motorists who drive into them. If so, the 'lacewood' timber from these beautiful trees is likely to end up as trays, bowls and ornaments.

ENGLISH OAK

Empire Builder

This tree, perhaps more than any other, is woven into the fabric of the English nation – its folklore, its traditions, its culture and its very history. Although not native only to England or indeed to the British Isles, it has become an English possession, its most adored tree. Its shape has become symbolic. It is a tree that most can identify, not just because of the unique shape of its leaves, but from its outline, even at a distance. For many, a visual delight is to see the winter silhouette of an oak's empty branches against a red sky at dawn or dusk. In the words of Rudyard Kipling: 'Of all the trees that grow so fair, Old England to adorn, Greater are none beneath the Sun, Than Oak and Ash and Thorn...'

The oak has played its part in England's history. More than 500 pubs bear the name Royal Oak, to mark the tree near Boscobel House in Shropshire in which the future King Charles II hid from the Roundheads after his defeat at the Battle of Worcester in 1651. The tree that stands there today is a 300-year-old descendant of the original, known as 'Son of Royal Oak'.

The Vikings arrived in England in oak ships, and 500 years later the royal forests, notably the New Forest in southern England, provided timber for British warships that enabled the exploration and colonization of much of the world, creating the largest empire in history. Some 600 oaks were felled to build Henry VIII's famous *Mary Rose*. It took 5,500 to build Nelson's flagship *HMS Victory*. As a result the forests, already stripped of some 4,000 oaks by the Great Storm of 1703, were severely depleted of mature specimens. However, there still exist some ancient trees with an upward-pointing arrow carved on the trunk, signifying that the tree was suitable for shipbuilding.

Many English oaks have lived beyond 1,000 years, and great efforts are made to protect such veterans. It is sometimes said that an oak tree takes 300 years to grow, 300 years to live and another 300 years to die. Even in death, left standing, they remain magnificent monuments.

OTHER COMMON NAMES
Common oak, pedunculate
oak, European oak

ORIGIN
Europe

CLIMATE AND HABITAT
Temperate and subtropical climates;
deep, moist soil; established trees
can tolerate short-term flooding

LONGEVITY
At least 800 years

SPEED OF GROWTH
20–50 centimetres/
8–20 inches per year

MAXIMUM HEIGHT
40 metres/130 feet

*The acorns of the English oak
are first produced when the tree is
between 25 and 40 years old.*

OTHER COMMON NAMES
Common sassafras, cinnamon
wood, filé gumbo

ORIGIN
Eastern North America

CLIMATE AND HABITAT
Temperate and subtropical
climates; open woodland; well-
drained, sandy loam soil

LONGEVITY
At least 150 years (through
suckering clones)

SPEED OF GROWTH
30–60 centimetres/
12–24 inches per year

MAXIMUM HEIGHT
25 metres/82 feet

*Sassafras leaves can form in
three distinct shapes, with one,
two or three lobes. The three-
lobed is the most common.*

Sassafras albidum (Lauraceae)

SASSAFRAS

The Beer Tree

Native to North America, in an area stretching from Ontario and Michigan to Florida and Texas, sassafras has been important to Native Americans for centuries. Its wood was used for dugout canoes and its leaves, roots and bark as food and medicine. The Choctaw are said to have been the first to use the dried, ground leaves as a seasoning and thickening agent in cooking, a habit that was later adopted by the Cajun and Creole cultures for such dishes as gumbo. The Cherokee used sassafras to heal wounds, banish fever, cure diarrhoea, ease rheumatism, cure colds and expel intestinal worms.

Whimsically, sassafras is said to have played a part in the discovery of the New World: Christopher Columbus may have picked up its sweet citrus scent above that of the briny ocean and followed his nose to find land. In the main, early European explorers and colonizers interacted peacefully with Native Americans, and began to learn many of the perceived attributes of sassafras. The English scholar and translator Thomas Harriot wrote of it in 1583 as 'A kind of wood most pleasant and sweet smell[ing], and of the most rare virtues in physic for the cure of many diseases'.

Such illustrious promotion increased the demand for sassafras on its introduction to England, which coincided with an outbreak of syphilis. During the early modern period it was generally agreed that God placed remedies for each disease in the lands where that particular disease flourished. Since syphilis was thought to have originated in North America, it followed that sassafras would be the remedy (it was later found to be completely ineffective). This demand, along with the use of the wood in construction and the culinary delights of the leaves, brought on the 'Great Sassafras Hunts' of the early seventeenth century. Inevitably, trade slowed as the natural population of sassafras diminished. It was subsequently popular as an ingredient in such beverages as root beer, but during the twentieth century it was discovered to be carcinogenic and must now be treated before use to remove the toxic safrole oil.

Ornamentally, the sassafras has much to offer. Its curious leaves, vibrantly red, yellow and orange in autumn, vary in shape from simple to three-lobed variations that are often described as 'mitten-shaped'.

Pinus parviflora (Pinaceae)

JAPANESE WHITE PINE

King of Bonsai

The pine is the seminal tree of Japanese gardens, and has long been appreciated in both Japan and China as a symbol of longevity, an understandably important quality since the previous record-holder of oldest inhabitant on the planet (died aged 117) was Japanese, as is the present record-holder (aged 116 at the time of writing).

Three main species are native to Japan: *Pinus thunbergii*, the Japanese black pine; *P. densiflora*, the Japanese red pine; and *P. parviflora*, the Japanese white pine. All three are revered in gardens, and the Japanese go to extraordinary trouble to shape the trees and present them in near perfect condition – while simultaneously retaining their natural environment-shaped canopies. Lower branches are allowed to remain, often carefully supported by props. The branches are manicured using a technique of crown-thinning, which involves the removal of some of the needle clusters to control the direction of growth and open the crown to allow light to reach all parts of the tree. This is bonsai on a giant scale, and it is occasionally referred to as cloud-pruning or *niwaki*.

P. parviflora is known in Japan as one of the Three Friends of Winter – pine, bamboo and plum. It is also the species that has found most favour as an ornamental outside Japan. Its short needles are decoratively twisted and often bluish in colour. Although the trunk of a mature tree can be 1 metre (3 feet) thick, this species is smaller and slower growing than the other Japanese pines, making it ideal material for bonsai. In Japan this pine is known as *goyomatsu*. An ancient specimen is to be found in the National Bonsai and Penjing Museum in Washington, DC. It was gifted to the United States by bonsai-master Masaru Yamaki in 1976 on the occasion of the US Bicentennial. His family had lived for generations just 3 kilometres (2 miles) from the epicentre of the atomic explosion in Hiroshima in 1945. Yamaki, his family and the bonsai survived, but he is said to have made no reference to this when he presented the ancient tree. The bonsai still lives, some 400 years old.

OTHER COMMON NAMES
Ulleungdo white pine, goyomatsu

ORIGIN
Japan, Korea

CLIMATE AND HABITAT
Prefers a cool climate, both
maritime and inland; very
hardy; tolerant of a wide
range of free-draining soils

LONGEVITY
At least 500 years

SPEED OF GROWTH
30–60 centimetres/
12–24 inches per year

MAXIMUM HEIGHT
25 metres/82 feet

*Japanese white pine leaves are
needle-like and borne in fives; the
seeds can be eaten raw or cooked.*

OTHER COMMON NAME
Wild plum

ORIGIN
United States, Canada

CLIMATE AND HABITAT
Warm to cold temperate regions;
prefers rich soil in full sun

LONGEVITY
Up to 100 years

SPEED OF GROWTH
30–45 centimetres/
12–18 inches per year

MAXIMUM HEIGHT
8 metres/26 feet

*The fruit of the American plum
is small with thick skin and sweet
flesh, and ranges in colour from
yellow to red to pink or purple.*

Prunus americana (Rosaceae)

AMERICAN PLUM

Cheyenne Treat

The wild plum, *Prunus americana*, is the most widespread among a number of species in the genus *Prunus* endemic to North America. The genus provides us not only with plums, but also with cherries (see page 58), apricots, almonds (page 60), peaches (page 28) and nectarines. A close relation of *P. americana* is *P. nigra*, which is known as the Canadian plum. Both species share the same habitat in Canada and the United States.

The first European settlers in North America valued the wild plum highly, and cultivated many varieties – some with red and some with yellow fruit, some sweet and others less so. The American plum is cultivated both for its fruit and as an ornamental tree, but is no longer grown in commercial orchards. Of the 200 forms previously selected for cultivation, few still exist, their place having been taken by varieties of the European plum, *P. domestica*.

The small, thorny, winter-hardy tree comfortably adapts to a variety of soils. The beautiful pure white flowers appear in abundance on its leafless twigs in early spring, opening wide to the sun. The tree attracts songbirds and animals as a safe environment. Deer munch on its bright green leaves, extracts of which are toxic to various insects.

Native Americans were cultivating American plums in their villages long before the arrival of European settlers, and orchards of plum trees were planted widely. The Pima of Arizona and Mexico and the Cheyenne were much attracted to the fruit, fresh and especially dried (a great treat), and also when cooked in desserts. It was also used to treat skin abrasions. Today, the fruit is mainly used to make preserves and jellies.

Although it can and does grow as a tree, the American plum is most often found as a thicket-forming shrub in hedgerows, and indeed one of the few remaining practical uses of *P. americana* is as a farmstead windbreak. When regularly cut or browsed, the tree develops a naturally suckering habit. Since it forms thickets, it is very useful for erosion control and is often used for restoration plantings.

Acer saccharum (Sapindaceae)

SUGAR MAPLE

Sap to Syrup

Acer saccharum, the sugar maple, is the archetypal maple, its foliage having the typical leaf shape as depicted on the Canadian flag. Native to Canada and northeastern parts of the United States, it is one of the best-loved trees: New York, West Virginia, Wisconsin and Vermont each claim the sugar maple as state tree. And *A. saccharum* is the most commercially important maple species in the world. All larger species are valuable for their timber, but the sugar maple, and its cousin the black maple (*A. nigrum*), are important for an entirely different reason.

With pecan nuts (see page 132) and blueberries, maple syrup is one of the great tastes of America. The indigenous peoples of North America demonstrated to the newly arrived European settlers how to make the syrup (as well as a crystalline sugar). Canada now produces some 70 per cent of the world's supply of maple syrup, and Vermont is the biggest supplier in the United States. Like rubber, maple syrup is made by tapping. The sap, which flows fastest from the tree during the spring snow melt, is heated in pans to evaporate excess water, leaving the sticky syrup. If you've ever wondered why maple syrup is expensive, well, 50 litres (11 gallons) of sap produces approximately 1 litre (just over 2 pints) of pure maple syrup.

Humans are not alone in their appetite for the sugar maple. The tree provides a feast for snowshoe hares, white-tailed deer, squirrels and even moose, which feed on the tree's buds, seeds, twigs and dark-green leaves. The syrup is not the only commercial reward from *A. saccharum*. It is valued for its timber, too. In 2001 the Major League baseball player Barry Bonds swapped his traditional ash bat for one made from maple. That season, he hit 73 home runs – a record.

But, sweetness aside, the sugar maple brings beauty, free of charge, each autumn (or fall). Its leaves turn from green to yellow, to red, and finally to shades of brown. Every year the display attracts thousands of tourists. And Johnny Mercer's lyric for Joseph Kosma's *Autumn Leaves* remains as popular now as it was when he wrote it more than 70 years ago.

OTHER COMMON NAME
Rock maple

ORIGIN
United States, Canada

CLIMATE AND HABITAT
Cool climate; rich, moist soil

LONGEVITY
Up to 400 years

SPEED OF GROWTH
30–60 centimetres/
12–24 inches per year

MAXIMUM HEIGHT
40 metres/130 feet

*The maple has a very distinctive
leaf shape, which is the national
symbol of Canada.*

In the commercial production of nutmeat, specially developed machines are used to extract the kernels from the fruit.

OTHER COMMON NAME
Eastern black walnut

ORIGIN
United States, Canada

CLIMATE AND HABITAT
Warm, temperate climates on fertile, mostly riparian soil

LONGEVITY
Up to 300 years

SPEED OF GROWTH
30 centimetres–1.2 metres/ 1–4 feet per year

MAXIMUM HEIGHT
40 metres/130 feet

Juglans nigra (Juglandaceae)

Black Walnut

North American Nutmeat

Like its cousin the English walnut (see page 36), *Juglans nigra* is valued for both its fruit and its wood. It is native to North America, where it has played an important part in the lives of the indigenous tribes for thousands of years. It is a majestic tree, tall and strong, with a spreading crown that provides welcome shade in the summer. It is also tolerant of cold and can be identified in leafless winter by the presence of a pith or spongy tissue inside its twigs. When buds emerge in the spring, they have at their base horseshoe-shaped leaf scars left by fallen leaves. The name black walnut comes from the dark hardwood and the distinctive, almost black, deeply fissured bark that is found even on young trees.

The tree's dark heartwood is especially beautiful. It is straight-grained, mottled, naturally durable and a joy to work with. The fruits are large – about the size of limes – and can cause damage to vehicles parked beneath the tree in autumn. Although they were valuable to Native Americans and early European settlers as a source of fat and protein, the kernels are notoriously difficult to extract whole from their shells. It takes strength and patience to complete the process, at the end of which the opener's hands will be stained black. These husks were once used to make a persistent yellowish-brown dye or ink.

The Native Americans have an unbelievable list of medicinal uses for the leaves, husk, bark and nuts of the tree: as a mosquito repellent, a skin ointment, a laxative and a cure for diarrhoea (an amazing dual treatment), for relief from fever and kidney ailments, to ease toothache, combat syphilis and cure snakebite.

It is thought that black walnut trees exude biochemical elements from their roots to harm competitors. This phenomenon, which also occurs in some other plants, is known as allelopathy (see also page 92). Walnuts are famous for being allelopathic, but Pliny the Elder was exaggerating when he wrote: 'The shadow of the walnut tree is poison to all plants within its compass.' The tree is not toxic to all plants, and certainly not to humans. These days walnut nutmeat is much used as a crunchy addition to baked goods, and in flavouring.

Cinchona officinalis (Rubiaceae)

QUININE

Tonic for the Troops

Cinchona officinalis is a shrub or small tree found in western South America, particularly in Ecuador, but also in Peru, Bolivia and Columbia, native to the slopes of the Andes. It is a plant of the moist montane tropics, growing at elevations from 1,500 to 2,700 metres (5,000 to 8,000 feet). The tree is evergreen, with red-pink blossom.

Cinchona trees remain the only economically viable source of the antimalarial drug quinine, which was first isolated from the bark of the tree in 1820 by the French chemists Pierre Pelletier and Joseph Caventou. Powdered bark extracts had been used to treat malaria as early as 1632, since the Jesuit brother and apothecary Agostino Salumbrino had observed the Quechua people of the Peruvian highlands adding the powdered bark to a sweet drink in order to reduce shivering. The Quechua's shivers were, in fact, caused not by malaria but by the cold, but coincidentally the quinine in the bark was to prove effective against the disease.

A couple of hundred years later malaria threatened the very heart of the British Empire. British soldiers, officials and their womenfolk faced an often mortal threat from the disease. The British began mixing the powdered bark of the tree with soda and sugar, creating the basis of medicinal tonic water. A British officer stationed in India is said to have mixed this tonic water with gin, already the favourite tipple of the ruling class. The ensuing G&T became the most popular way of mixing medicine with pleasure. Winston Churchill famously declared: 'The gin and tonic has saved more Englishmen's lives, and minds, than all the doctors in the Empire.' Quinine (not in gin and tonic form) was also crucial in reducing the horrendous death rate among labourers building the Panama Canal in the early twentieth century.

During World War II there were fears that the bark would become inaccessible, and this, coupled with increasing demand for quinine, made it necessary to find a synthetic substitute. This was at last produced by American chemists in 1944. Others have been synthesized since then, but none has competed in economic terms with the natural version, or provided any greater benefits.

OTHER COMMON NAMES
Jesuit's bark, Peruvian bark

ORIGIN
South America

CLIMATE AND HABITAT
Wet tropical climate; fertile soil

LONGEVITY
Unrecorded. Destructively
harvested at 8–12 years old

SPEED OF GROWTH
1–2 metres/3–6 feet per year

MAXIMUM HEIGHT
8 metres/26 feet

*Quinine has the clusters of white or
pinkish flowers typical of members of the
Rubiaceae family, such as the coffee tree.*

OTHER COMMON NAMES
Juniper, jenever

ORIGIN
North America, Europe,
Asia, North Africa

CLIMATE AND HABITAT
Cool to warm temperate and
subtropical climates; adaptable,
but grows best on chalky soil

LONGEVITY
Up to 600 years

SPEED OF GROWTH
2–5 centimetres/
1–2 inches per year

MAXIMUM HEIGHT
15 metres/50 feet

*The small, bitter juniper berry is used
to flavour both gin and genever.*

Juniperus communis (Cupressaceae)

COMMON JUNIPER
Mother's Ruin

Perhaps the most remarkable feature of the common juniper is its geography: it has the largest natural range of any woody plant species, with a circumpolar distribution across the northern hemisphere, being found in Europe (including the UK), North Africa, Asia and North America. It is also one of the most variable species; often seen as a shrub of varying shape, it is capable of growing to tree-like proportions. The variation is mostly genetic, but it is also caused by environmental factors, including climate, geography and animal predation. The tree has flaky, fibrous, grey-brown bark, white sapwood and brownish-pink heartwood. In some ways it resembles a lanky gorse bush, its trunk twisting and curving, and its leaves clustering at the ends of its branches.

The Dutch for juniper is *genever* or *jenever*, and this gives a clue to the most famous culinary use of the bitter blue fruit it produces. Genever – sometimes known as Hollands – and gin are now different spirits, but they share a common heritage and ingredient in juniper berries. London dry gin was popularized by the Dutch king William of Orange, who became King of England, Scotland and Ireland in 1689. William and his queen, Mary, effectively deregulated distillation, making it possible for almost anyone to produce gin.

In the eighteenth century the fruit of *Juniperus communis* was almost the ruin of the working classes, who were sometimes paid their wages in gin. In a single year, 1742, the sale of gin in London alone rose to just over 7 million gallons (with a population of just over 750,000), ushering in the era of extreme drunkenness and civil unrest known as the 'Gin Craze' that was graphically portrayed by the artist William Hogarth in his painting *Gin Lane*. It wasn't until the early nineteenth century, however, that gin became known as 'mother's ruin'. It is now a fashionable drink again, normally served with ice, a slice of lemon and a little *Cinchona officinalis* (see page 148).

Acer campestre (Sapindaceae)

FIELD MAPLE

Fit for a Luthier

The field maple is one of the loveliest of wild trees, with its distinctive corky bark, lime-green spring leaves and flowers, and butter-yellow autumn colour. Left alone, it is a medium-sized, slow-growing, deciduous tree distributed widely throughout Europe, and also to be found growing wild in the Atlas Mountains of North Africa and in South West Asia. Closely related to the sycamore, it is the only species of maple that is truly native to the British Isles. There it is more often seen as a frequently lopped, stunted hedgerow plant than as a tree, which is why it is called the 'field' maple. The maple grows far too slowly to be of much commercial use as timber, but that same lack of speed, with its small leaves in a variety of colours, makes it ideal as a bonsai specimen or simply as a garden tree.

The wood of the field maple does have one great claim to fame. In the seventeenth and eighteenth centuries Antonio Stradivari lived in the Italian city of Cremona, where he made the most famous stringed musical instruments in history. Stradivari used the field maple's tonewood – wood that has special tonal properties – for his violins. They are still widely regarded as superior to any other violins produced since. For some 300 years it has remained a great mystery as to why apparently similar instruments have not been judged at least equal to these now heavily repaired Cremonese antiques. Numerous theories attempting to explain the superiority of a Strad have failed to do so scientifically. One theory has it that Stradivari, along with his neighbour and fellow luthier – a maker or repairer of stringed instruments – Giuseppe Guarneri, used chemicals to cure the wood before making the instruments. A more plausible explanation may be that field maple trees felled between about 1640 and 1750 (roughly the span of Stradivari's life) were subject to much cooler temperatures than those of today. The master luthiers were working in the period known as the Little Ice Age. The trees would have grown even more slowly during this time, so their growth rings would have been narrower, and their wood correspondingly denser, than field maples cultivated since.

More humbly, *masern*, the Welsh word for maple, has given its name to *mazer*, a small turned bowl of maple ornamented with silver.

OTHER COMMON NAME
Hedge maple

ORIGIN
Europe, South East Asia

CLIMATE AND HABITAT
Temperate and subtropical
climates where it is found in a
wide range of habitats excluding
wetland; often found in hedging

LONGEVITY
Up to 300 years

SPEED OF GROWTH
30–50 centimetres/
12–20 inches per year

MAXIMUM HEIGHT
25 metres/82 feet

*Field maple leaves are eaten by the
caterpillars of several species of
moth, including the sycamore moth.*

OTHER COMMON NAMES
Chile pine, Chilean pine

ORIGIN
Chile, Argentina

CLIMATE AND HABITAT
Temperate montane; free-draining, acidic, volcanic soil

LONGEVITY
Up to 1,800 years

SPEED OF GROWTH
15–60 centimetres/
6–24 inches per year

MAXIMUM HEIGHT
45 metres/148 feet

The monkey puzzle tree has been around for 20 million years, so its sharp, hard leaves would once have protected it against dinosaurs.

Araucaria araucana (Araucariaceae)

MONKEY PUZZLE

Seeds from the Past

It is comparatively easy to recognize the monkey puzzle from any other tree. Its shape is unique – not unlike an artificial Christmas tree – and it has very thick leaves with sharp edges and points. The origin of the popular name is derived from its cultivation in England in the mid-nineteenth century; the politician Sir William Molesworth, the proud owner of a tree, was showing it to a group of friends on his estate in Cornwall when one of them remarked, 'It would puzzle a monkey to climb that.'

Araucarias are a very primitive type of conifer, often referred to as fossil trees. The Wollemi pine (see page 11), recently discovered near Sydney, Australia, is closely related. Their shared ancestry dates back to a time when Australia, Antarctica and South America were joined as the supercontinent known as Gondwanaland, which existed until about 320 million years ago. The name of *Araucaria araucana* is derived from the Araucanos, a group of tribes, some of whom still live in Chile and Argentina. They consider this tree to be sacred, and its seeds are an important food to them, eaten raw or roasted. The dried seeds are made into flour, which the Araucanos use to make a mildly alcoholic fermented holy drink, or *muday*.

Towards the end of the eighteenth century Captain George Vancouver circumnavigated the globe in command of HMS *Discovery*, named in honour of Cook's ship. The voyage came to be known as the Vancouver Expedition and, with one or two exceptions, put the northwest coast of the Americas on the map. The ship's naturalist was the Scottish physician and accomplished botanist Archibald Menzies, who is commemorated in the Latin name of the Douglas fir (*Pseudotsuga menziesii*; see page 172). During the voyage, while dining with the governor of Chile, Menzies was served the seeds of a conifer for dessert. Keeping some aside, he sowed them when back on board ship. He returned to England in 1795 with five healthy plants, one of which was planted at the Royal Botanic Gardens at Kew.

There are many myths about the tree. It is said that talking while passing under the tree will bring bad luck, and that planting the tree on the edge of a graveyard will ensure that the Devil does not gatecrash a funeral. The Devil is also said to live in a monkey puzzle tree, since Satan apparently possesses the skills of a monkey.

Pinus contorta subsp. *latifolia* (Pinaceae)

LODGEPOLE PINE
Timber for the Tipi

As the last few pages of this book were being written, forest wildfires raged in California, on an unprecedented scale. More than 750,000 hectares (1.9 million acres) of forest were lost. Devastating as this may sound, it is part of the natural ecology of conifer forests – fires cleaning up old wood from the forest floor and making a new seedbed for a new generation of trees and plants to grow. Cones of many types of pine will open to distribute the seeds only after a fire, and the lodgepole pine is a classic example of a tree that has evolved to take full advantage of fire.

Native to a vast area of western North America, this tree is common in the Rocky Mountains, in subalpine regions from 100 to 3,500 metres (330–11,500 feet) above sea level. Unlike many of its native competitors, the lodgepole tree has evolved to grow in problematic surroundings: in soil low in nutrients or saturated and highly acidic, or in intense heat such as that in the geothermal areas of Yellowstone National Park. Where none of these conditions prevail, the lodgepole generates a suitable environment, thanks to its thin bark, its resinous sap and its ability to grow in dense forests. A lodgepole fire – such as that of 1988, when over 3,000 square kilometres (740,000 acres) of Yellowstone burned after a lightning strike – is particularly devastating and fierce, and the heat destroys not only the trees and timber above ground, but also the soil ecology below. Nothing much woody grows after a lodgepole fire – except lodgepole – but some wild flowers and grasses have evolved to take advantage, and it is truly remarkable how quickly life returns after the cooling off.

The lodgepole pine is named for its common use as the structural poles for the Native American tipi shelters that were common across the Great Plains. Tribes made extended journeys across the plains to obtain the long, straight timbers, which grew only in the mountains. This pine continues to provide highly desirable pole timber, which today is used for post-and-rail fences and for structures such as log cabins and barns.

OTHER COMMON NAME
Doghair pine

ORIGIN
United States, Canada

CLIMATE AND HABITAT
Temperate montane subalpine
forest; waterlogged to sandy soils

LONGEVITY
Oldest recorded at 630 years

SPEED OF GROWTH
20–90 centimetres/
8–35 inches per year

MAXIMUM HEIGHT
40 metres/130 feet

*Lodgepole pine cones contain a
resin between their scales that
breaks only when temperatures
reach 45–60°C (113–40°F).*

OTHER COMMON NAMES
American persimmon, eastern
persimmon, sugar-plum, simmons

ORIGIN
Southeastern North America

CLIMATE AND HABITAT
Subtropical and warm temperate
climates; free-draining, rich soil

LONGEVITY
Up to 150 years

SPEED OF GROWTH
10–60 centimetres/
4–24 inches per year

MAXIMUM HEIGHT
20 metres/66 feet

*The bitter persimmon fruit bears
a crown at its apex, the remains of
the flower part known as the calyx.*

Diospyros virginiana (Ebenaceae)

PERSIMMON

A True 'Wood'

Diospyros virginiana, the persimmon, is a deciduous tree that is native to the southeastern United States, although it has grown as far north as Connecticut. In winter it is one of the easiest trees to identify, due to the thick, rectangular dark-grey blocks of its bark. The genus *Diospyros* also grows successfully in southern Europe, where the fruit of a related species, *D. kaki* (kaki or Japanese persimmon), was known to the ancient Greeks as 'the fruit of the gods', hence *Diospyros* – *dios* meaning divine and *spyros* meaning wheat or grain.

Since prehistoric times, the persimmon tree has been cultivated by Native Americans for its fruit and its wood. Its orange globe-shaped berry is a little smaller than a tennis or baseball and at its apex bears a distinct crown, the remaining flower calyx (the outer covering of the flower while in bud). Today the fruit is used in pies, puddings, sweets, syrups, jellies and ice creams. Southerners in the United States like to induce the unwary to taste the unripe fruit, which is almost painfully astringent. This is caused by the presence of tannin in the young fruit, similar to that in *Cinchona* or quinine (see page 148). The mature ripe fruit is rich in Vitamin C and is eaten cooked, raw or dried.

Persimmon wood is sometimes known as white ebony because, unlike other species, its heartwood rarely turns dark. It is highly regarded by woodturners for its great hardness and unpredictable pattern. It is the finest wood for billiard cues, the heads of golf clubs and shoe lasts. In the days before metal-heads, the first shot off the tee on a golf course, on a long hole, was always taken with what were called 'woods'. In Scotland, the timber used to make golf clubs changed from local timbers from 1900 onwards, when hickory (see page 160) and persimmon were imported from the United States – hickory for the shafts, persimmon for the heads.

Carya ovata (Juglandaceae)

SHAGBARK HICKORY

Hard as Hickory

The seventh President of the United States, Andrew Jackson (in office 1829–37), was popularly known as Old Hickory, comparing the toughness of the wood with the man who was prepared to take on the Bank of America, and win. The Hermitage in Tennessee was Jackson's home and, long before his death, he planned his grave site, planting it with a variety of trees including six shagbark hickories. The trees stood until a storm demolished them in 1998.

The shagbark hickory, *Carya ovata*, is the most common hickory species, native to the eastern United States and southeastern Canada. Mature specimens of this large deciduous tree are easy to recognize because, as their name suggests, the mature tree has remarkably shaggy bark. The bark of young specimens, on the other hand, is surprisingly smooth. The hickories are closely related to pecans (see page 132) and walnuts (see pages 36 and 146), and there are clear similarities in their nuts.

The nutritious nuts of the hickory were an important food for Native Americans, but unfortunately they are unsuited to commercial culture owing to the long time it takes for a tree to produce a sizeable crop, and the species' unpredictable seed set. Trees begin producing seeds when they are about 10 years old, but large quantities are set only after about 40 years. Nut production is not annual, and good crops may appear only every three to five years. The entire crop may then be lost to animal predation. The pecan is, therefore, the hickory of choice for its fruit, although an extract tapped from the shagbark hickory is used to make an edible syrup similar to maple syrup, slightly bitter and smoky in taste.

Like walnut, hickory wood is hard, dense, heavy and shock-resistant, and makes a wonderful smoke for curing and cooking meat. It is also used for tool handles, bows, cartwheel spokes, drumsticks and, once upon a time, golf-club shafts (which used to be called 'hickory sticks'). Baseball bats were once made of hickory, but are now more commonly made of ash (see page 44), which is considerably lighter. Hickory's combination of strength, hardness and resilience is not found in any other wood.

Other common name
Hickory

Origin
Eastern North America, Canada

Climate and habitat
Subarctic and temperate climates; rich, free-draining soil

Longevity
Up to 350 years

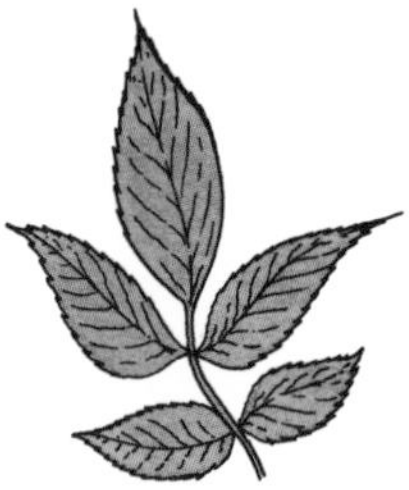

The shagbark hickory leaf turns golden-brown in autumn.

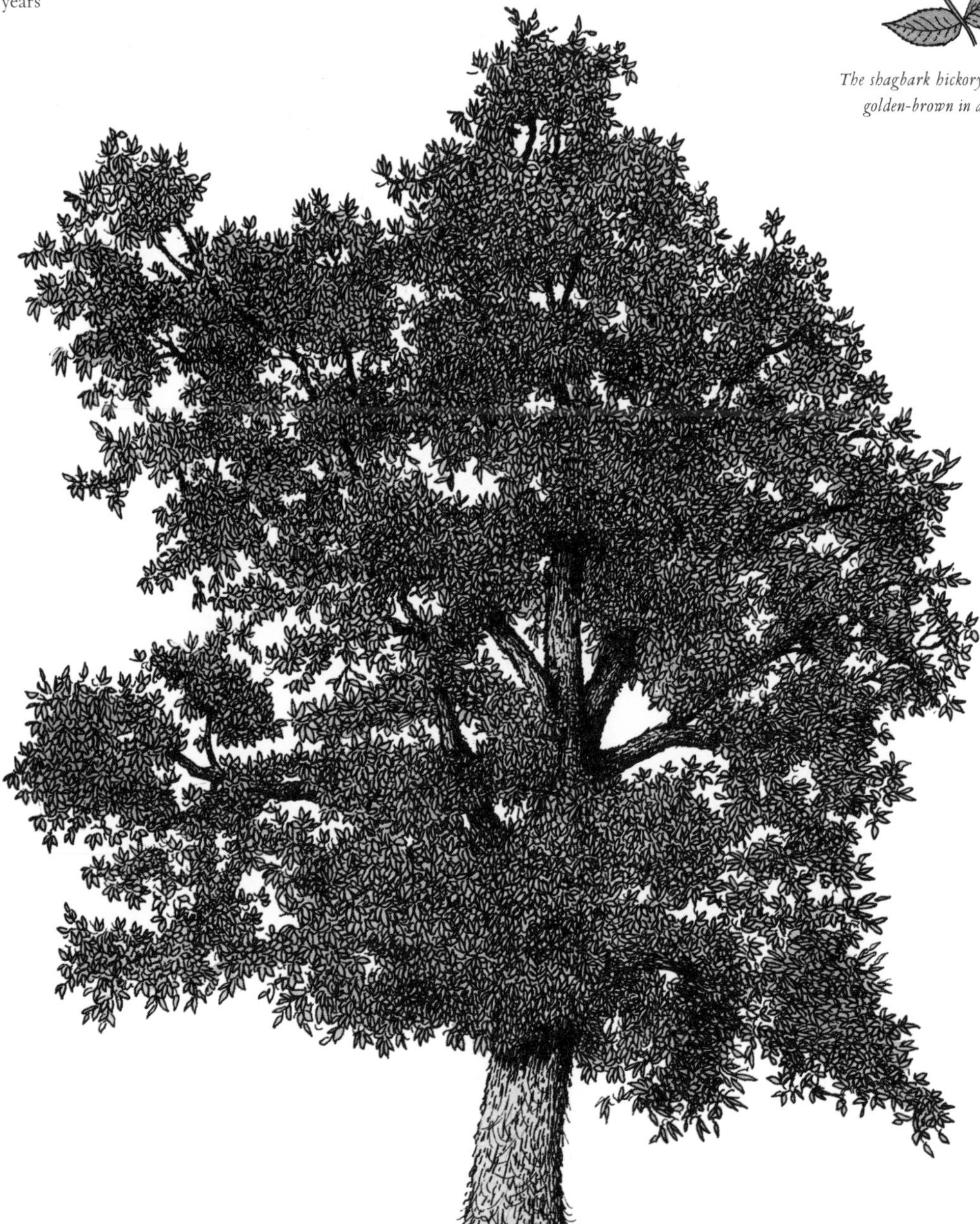

Speed of growth
30–45 centimetres/
12–18 inches per year

Maximum height
35 metres/115 feet

Mahogany is sometimes known as big-leaf mahogany because of the size of its pinnate leaves.

OTHER COMMON NAMES
Big-leaf mahogany, genuine mahogany, West Indian mahogany

ORIGIN
Mexico, Central and South America, West Indies

CLIMATE AND HABITAT
Rainforest; often towers above competing trees; deep, free-draining, moist to dry, sandy loam soils

LONGEVITY
Up to 350 years

SPEED OF GROWTH
50 centimetres–2 metres/ 1½–6½ feet per year

MAXIMUM HEIGHT
60 metres/200 feet

Swietenia macrophylla (Meliaceae)

MAHOGANY

Fine Furniture

Perhaps more than any other wood, mahogany conjures up images of the finest furniture – an elegant chair, a beautiful table, an exquisite bureau. In many ways, mahogany is the gold of timber. For some time, the Spanish kept it as their New World monopoly, but by the early eighteenth century France and England wanted a well-polished slice of the action. In 1721 the timber, brought from the Caribbean to England, was deemed so important to British trade that import duty was abolished by the Naval Stores Act. By the mid-eighteenth century 500 tons of the timber were imported each year; 30 years later this had soared to 30,000 tons. The Act also had the effect of increasing the export of mahogany to the 13 British colonies of North America.

Originally designated joinery-grade timber, mahogany soon became the choice for furniture-making. It was big business, and native forests were exploited mercilessly. Today the timber is more likely to be farmed on large plantations.

Genuine mahogany comes from only three recognized species of the genus *Swietenia*. Other tree genera produce similar timber, but they cannot be called 'genuine'. True mahogany is native to an area reaching southwards from southernmost Mexico, through Central America to tropical South America. In some ways the tree is not dissimilar to the ash when cultivated away from rainforest competition, being of similar proportions and with a composite pinnate leaf and rounded, spreading crown. The rough-barked trunk is distinct, not only far darker, but also sporting a wide, often contorted buttress, its roots forming architectural feet that support the lofty trunk.

Mahogany timber is straight-grained and has no knots or air pockets. Its durability and resistance to rot make it especially valuable for high-grade woodwork. These qualities, coupled with its beautiful reddish-brown colour, which polishes to a rich red sheen, make *Swietenia macrophylla* the mahogany king of the showroom.

Araucaria heterophylla (Araucariaceae)

NORFOLK ISLAND PINE

Captain Cook's Treasure

Next time you holiday in the Mediterranean, look out for the Norfolk Island pine: it will be one of the taller trees on the skyline, perhaps in the company of *Cupressus sempervirens* (see page 70). As its name suggests, this ancient conifer is native to Norfolk Island, a small island north of New Zealand. Despite the fact that the early Polynesians had known of its existence thousands of years ago, Captain James Cook is generally credited with its discovery in 1774, on his second voyage to the South Pacific, this time aboard HMS *Resolution*. Cook had observed the trees while still some distance from the island and believed their trunks to be tall and straight, and therefore with good potential to serve as ships' masts. Unfortunately, they did not live up to expectation as the timber was found not to be resilient enough.

It is not a true pine, but instead belongs to a family of ancient coniferous trees, a group that includes the monkey puzzle tree (see page 154), the bunya pine (see page 188) and the Wollemi pine, *Wollemia nobilis*, which was discovered in a remote canyon 150 kilometres (93 miles) from Sydney, Australia (see page 11). These fossil trees, which were widespread in the Jurassic and Cretaceous periods, now exist only in disjointed populations in the southern hemisphere.

The Norfolk Island pine's unique appearance has made it one of the most commonly planted ornamental conifers in warmer parts of the world, especially maritime areas, both north and south of the Equator. It has a symmetrical triangular outline, with inclined branches, each with foliage that radiates like the bones of a ribcage. It is planted in avenues as well as individually, and in many countries it is cultivated as a houseplant and often as a Christmas tree. It is easiest and best grown from seed. Unusually, if cuttings are rooted from its lateral branches they will continue to grow horizontally – never producing an erect stem. The tree really needs a frost-free climate if it is to grow well, although it will tolerate short-term freezing.

OTHER COMMON NAMES
Star pine, Polynesian pine

ORIGIN
Norfolk Island (South Pacific)

CLIMATE AND HABITAT
An adaptable tree for mild
climates, but most at home in
maritime climates, where it
is very tolerant of salt; well-
drained, preferably sandy soil

LONGEVITY
At least 170 years

SPEED OF GROWTH
30–60 centimetres/
12–24 inches per year

MAXIMUM HEIGHT
65 metres/213 feet

*The Norfolk Island pine is known
to purify the air of harmful
compounds, such as those found in
paints, cleaning agents and glues.*

When uncooked, the seeds are toxic
to humans and livestock; if cooked
they can produce a type of coffee.

OTHER COMMON NAME
Chicot

ORIGIN
United States, Canada

CLIMATE AND HABITAT
A climate that provides plenty of
summer heat; adaptable hardy tree
preferring rich, moist or wet soil

LONGEVITY
Up to 150 years

SPEED OF GROWTH
30–60 centimetres/
12–24 inches per year

MAXIMUM HEIGHT
30 metres/98 feet

Gymnocladus dioica (Fabaceae)

KENTUCKY COFFEETREE
A Mammoth Bean

When the first white settlers arrived in Kentucky from the Atlantic states, they were introduced by the Meskwaki Native Americans to a tree now known as the Kentucky coffeetree. The Meskwaki had fared badly at the hands of both the French (in the mid-eighteenth century) and the Americans (after the War of Independence), but they passed on their practices of roasting and grinding the seeds of the trees to make a substitute for coffee.

Gymnocladus dioica has a fascinating history. In prehistoric times it owed its survival to mammoths and mastodons long before the Meskwaki came along. About 5 million years ago the tree's seed pods were food for the giant herbivores that were dominant at that time, and that was how the tree evolved and spread. The incredibly hard black seeds are about the size of a shelled Brazil nut. They will not germinate unless they are soaked in acid for a time to digest the seed coat. The creatures whose stomachs would have done this are long extinct, but the tree survives in wetlands, where its seeds can lie buried until the seed coat has rotted sufficiently. Gardeners and breeders trying to propagate from seed can choose between using concentrated sulphuric acid for a few hours – with the consequent environmental problems – or filing off each seed coat individually, a labour-intensive task.

Now valued as an ornamental, the Kentucky coffeetree is unusual in being one of the last trees to break dormancy in the spring and one of the first to drop its foliage in the autumn. The generic name *Gymnocladus* comes from the Greek for 'naked branch', referring to the stout, stub-like branches that are still devoid of leaves when most other trees are well clothed. The foliage as it emerges is very attractive, the dark-blue-and-green leaves elegant and bipinnate (each one being a composite of smaller leaves, themselves subdivided, much like many ferns). The tree's short period in leaf makes it ideal for creating urban shade when necessary, while giving as much room as possible for winter sunshine.

Quillaja saponaria (Quillajaceae)

SOAP BARK TREE

Nature's Cleaner

The soap bark tree is an evergreen, native to the warmer, temperate parts of Andean Chile and Peru, where it flourishes on land up to 2,000 metres (6,500 feet) above sea level. It was originally considered to be a member of the Rosaceae (rose) family, but has now been reclassified as belonging to the family Quillajaceae, the only other member of which is *Quillaja brasiliensis*, a native of neighbouring Brazil.

The soap bark tree is so called because it produces saponins – lathering, soap-like phytochemicals – from its inner bark. Their use in cosmetics is growing, under the pressure of environmental concerns, but also because of the saponins' unique properties. Soap bark saponins can act as natural foaming agents to offer a real alternative to the more commonly used synthetic surfactants (surface-acting substances). In the tree's native lands saponins from the bark are used for washing clothes and as shampoo, when mixed with alcohol and essence of the bergamot orange. In its uses, the tree is almost a cross between pharmacy and beauty parlour, for the bark is also valued by the Andean people as an expectorant, to loosen chesty coughs.

The slow-growing tree is narrowly upright, with small, shiny, short-stalked grey-green leaves. It produces abundant sprays of small creamy-white flowers that attract many pollinating insects, and which are followed in autumn and winter by small brown fruits. These fruits split open to release between 10 and 20 tiny winged seeds. The outer bark of the tree is thick, dark and very tough. Altogether, *Q. saponaria* has a quiet charm and a rugged ability to survive. It is often used for reforestation in arid soils, and could also be used to a greater extent in drought-stricken parts of the world. It has been employed successfully in California, particularly in San Francisco, where its toleration of pollution, its evergreen foliage and its straight-backed stance combined with gracefully arching branches contribute a great deal to the urban landscape.

ORIGIN
Chile, Peru

CLIMATE AND HABITAT
At home in warm temperate or
Mediterranean climates, where
it will tolerate a few degrees of
frost but no long-term wet; free-
draining neutral to acid soils

LONGEVITY
Unknown (a very slow-
growing species, potentially
capable of living to a great
age, but so far unrecorded)

SPEED OF GROWTH
15–30 centimetres/
6–12 inches per year

MAXIMUM HEIGHT
20 metres/66 feet

*Attractive white flowers in the form of
stars burst forth from the soap bark tree
in late spring and last into summer.*

OTHER COMMON NAMES
Yellow poplar, canoewood, tulip
poplar, American tuliptree

ORIGIN
United States, Canada

CLIMATE AND HABITAT
Cold temperate climate; rich,
moist soil; very hardy

LONGEVITY
Up to 350 years

SPEED OF GROWTH
20–60 centimetres/
8–24 inches per year

MAXIMUM HEIGHT
60 metres/200 feet

*Tulip-tree flowers first appear when
the tree is about 15 years old; in
autumn they turn buttery yellow.*

Liriodendron tulipifera (Magnoliaceae)

Tulip Tree

American Canoewood

The tulip tree – also sometimes known as 'the tulip poplar' because of the similarity of its timber to true poplars – has for centuries been planted as an ornamental 'feature' tree in parks, gardens and arboreta all over the world. Closely related to the magnolia, it has been around for a very long time: some 50 million years or more in America, and 100 million years elsewhere. Amazingly, more than one species has survived. While *Liriodendron tulipifera* thrives in the eastern United States, the very similar *L. chinense* is happy in China. The two species are the ancient legacy of a world that was at one time joined up.

In North America, *L. tulipifera* is not only the tallest species of hardwood tree, but also the second highest tree of all, second only to conifers such as coast redwood. It is one of the fastest-growing hardwood trees on the American continent, producing a pale timber that is easy to work, durable – an unusual quality in a fast-growing tree – light in weight and usually yellowish-green with pinkish parallel grooves. Numerous common names have been given to *L. tulipifera* at different times by different peoples and for its various uses. Eastern Native Americans referred to it as 'canoewood', since it was a choice material for their dugout canoes.

The tulip tree is perhaps most easily identified by the bite that seems to have been taken from each leaf. The flowers that give the tree its name bloom in late spring, yellowish-green with a small flash of orange. But the tree takes its time to display blossom, often blossoming for the first time in its mid-teens, and then sparsely, hiding the beauty of the flower. The fruits that follow are cone-shaped.

The first President of the United States, George Washington, loved trees and was a passionate gardener. His estate at Mount Vernon, Virginia, became a personal arboretum, of which only four original trees survive. Two of them are tulip trees planted in 1785, one year after his return to his beloved home after the War of Independence and its immediate aftermath. The giant of these survivors is 43 metres (140 feet) tall.

Pseudotsuga menziesii (Pinaceae)

DOUGLAS FIR

The Christmas Tree

The Douglas fir – not a true fir – is a giant among conifers, and trees in general.
For a while, a Douglas fir felled on Vancouver Island in 1895 held the record
as the world's biggest tree – 127 metres (417 feet) in felled length. Nowadays,
the Douglas fir is ranked third in the world in terms of height, after the coast
redwood (*Sequoia sempervirens*) and the Australian mountain ash (*Eucalyptus
regnans*). Native to North America, *Pseudotsuga menziesii* was introduced to
Britain by the Scottish naturalist David Douglas in 1827. Douglas sent seeds of
the tree to Europe, where it has been widely cultivated for its timber ever since.
The tree therefore takes its common name from one worthy Scot, but gets its
scientific name from another: the physician, plant-hunter and naturalist, the
man who discovered and introduced the monkey puzzle tree (see page 154) and
who was also Douglas's greatest rival: Archibald Menzies.

This mighty tree is now cultivated throughout the world, is one of the
most important forestry trees for the construction industry and is also valuable
for boat-building, its tall, straight trunk being particularly suitable for masts.
Its timber is hard but flexible, with few knots. The tree can live for a thousand
years or more, and because of this, deadwood cavities appear in the trunk,
making nesting sites for birds of prey such as buzzards, sparrowhawks and red
kites. A species of moth feeds on the leaves, and the seeds are eaten by finches
and small mammals. In Scotland, the Douglas fir is home to the pine marten
and the red squirrel, both of which have long, sharp claws that enable them to
climb and live happily in the tree. Native American myth suggests that mice
hide in the cones of the Douglas fir to escape forest fires.

In the United States over the last 90 years, the Douglas fir has become one
of the most popular species for Christmas trees. Originally, these trees were
harvested from the wild, but thankfully since the 1980s the specimens grown
for this intensely seasonal use are now farmed on plantations, where they are
managed on a harvesting cycle of between seven and ten years.

OTHER COMMON NAME
Oregon-pine

ORIGIN
Pacific coast of North America,
from British Columbia to California

CLIMATE AND HABITAT
Best suited to maritime climate,
but adaptable to a range of
growing conditions; prefers
moist, neutral to acid soils

LONGEVITY
Typically up to 650 years;
oldest recorded at 1,200
years from ring counts

SPEED OF GROWTH
20–60 centimetres/
8–24 inches per year

MAXIMUM HEIGHT
100 metres/330 feet

*The Douglas fir is not a 'true' fir.
True fir trees have cones that sit
upright on their branches, while the
Douglas fir has cones that hang down.*

Wind carries pollen from male to female Sitka spruce cones. Three years later the female cones release their seeds.

OTHER COMMON NAME
Giant spruce

ORIGIN
Western coastal Canada and United States Pacific Northwest

CLIMATE AND HABITAT
Cool, moist maritime climate; very wet soils in cold temperate regions

LONGEVITY
500–700 years

SPEED OF GROWTH
10–60 centimetres/
4–24 inches per year

MAXIMUM HEIGHT
100 metres/330 feet

Picea sitchensis (Pinaceae)

SITKA SPRUCE
Fit for Flight

In previous centuries timber was required in enormous quantities for ship- and boat-building and for the construction of the railways, but at the turn of the last century it became required for an entirely new mode of transport: air travel. The Sitka spruce, sometimes known as the 'giant spruce', was the timber of choice for the Wright Brothers' Flyer, which achieved the first heavier-than-air controlled flight in 1903. Forty years later, during World War II, spars of Sitka spruce were used in the manufacture of the largely wooden plane, the de Havilland DH.98 Mosquito, nicknamed the 'Wooden Wonder', with a top speed of 378 mph (604 kph). One of the beauties of this enormous tree is its ability to grow tall and straight, so that there are few knots in the timber. It is very strong but also flexible, and the qualities that made it so good for sailing-boat spars made it a natural choice for the same in aircraft.

Like the Douglas fir (see page 172), the Sitka spruce is native to the west coast of North America, from Alaska in the north to the shores of northern California. After centuries of intensive logging, these native spruce forests are a fraction of what they once were, and most of the largest native forests are gone. However, this species has been widely planted for forestry elsewhere, particularly in Norway. The bark is a mixture of purple and grey, with plates forming as the tree ages. Its needle-like leaves are particularly sharp; the male flowers are soft and yellow; the female flowers are red, but are rarely visible since they grow at the top of the tree.

The Sitka spruce, the world's third-tallest conifer, was another of David Douglas's discoveries. It was named after the southern Alaskan community of Sitka on Baranof Island, where it can still be found. It lives long and grows quickly, producing up to 1 cubic metre (more than 35 cubic feet) of wood each year. Introduced in Britain in 1831, it is still one of the most popular trees for general forestry. Its wood is used mostly in boat and ship construction, pallets, packing boxes and pulp for high-quality paper.

Musa acuminata (Musaceae)

BANANA

Jamaican Delicacy

Musa is the genus of one of the earliest domesticated plants, the banana or plantain. Various estimates put the start of its cultivation at between 8000 and 5000 BC. *M. acuminata* is the main species that has contributed its genes to the modern banana. We now know through genetic fingerprinting that bananas are a hybrid strain, involving this species, which originated in the biogeographical region of Malesia (comprising Malaysia, Indonesia, New Guinea, the Philippines and Brunei), and *M. balbisiana*, a hardier and more drought-tolerant species from southern China. *Musa* is thought to have been named in honour of the physician to the Roman Emperor Augustus, Antonius Musa, who cultivated the exotic fruit between 63 and 14 BC.

The fruit has been cultivated since ancient times, preceding the cultivation of rice. Portuguese sailors first brought bananas to Europe in the early fifteenth century. It was grown in the Canary Islands and taken to the West Indies, finally reaching South America in the sixteenth century with the influential Spanish missionary and Bishop of Panama, Fray Tomás de Berlanga. Today's sweet banana was discovered in a Jamaican plantain plantation in 1836 by the farmer Jean François Poujot, who found a mutant plant with a sweet yellow fruit. Until that time the banana had been eaten only as a cooked vegetable. The sweet banana spread across the Caribbean to North America. Initially it was treated as a delicacy, to be eaten solely with a knife and fork. Today it is gobbled straight from its skin, an essential fast supply of energy for all, but especially top tennis players in between sets.

Bananas are, in fact, not true trees. The strength of their 'trunks' comes from tightly curled leaves which form a pseudo-stem. Bananas are the largest evergreen perennials on the planet, growing from giant corms below ground. Of its several species, the most remarkable is the giant banana, *M. ingens*, which was discovered in 1954 in the highland forests of Papua New Guinea. It grows to 15 metres (50 feet) tall, with a stem circumference of 2 metres (6½ feet) at the base, and its huge leaves are up to 5 metres (16 feet) long. At the other end of the scale is the Japanese banana, *M. basjoo*, which is small and hardy enough to be grown outdoors in temperate areas, such as the United Kingdom. It makes a superb, exotic-looking herbaceous perennial.

OTHER COMMON NAME
Plantain

ORIGIN
South East Asia

CLIMATE AND HABITAT
Wet tropical climate; fertile soil

LONGEVITY
Fruiting growth is 10–15
months, replaced by new
growth from the base

SPEED OF GROWTH
2–3 metres/6½–10 feet per year

MAXIMUM HEIGHT
6 metres/20 feet

*The banana is botanically a berry
of a large herbaceous flowering
plant – technically a gigantic herb.*

OTHER COMMON NAME
Silver willow

ORIGIN
Europe, Asia

CLIMATE AND HABITAT
Enjoys lowland moisture; often
found on riverbanks and around
lakes; wet and mostly acidic soils

LONGEVITY
50–70 years

SPEED OF GROWTH
60 centimetres–1.8 metres/
2–6 feet per year

MAXIMUM HEIGHT
30 metres/98 feet

*The caterpillars of several moth
species feed on the leaves of the white
willow, including the eyed hawkmoth.*

Salix alba (Salicaceae)

WHITE WILLOW

Leather on Willow

The white willow, so named for its silvery leaves with white undersides that are revealed in the lightest breeze, is native to Britain and mainland Europe, extending to western Asia. A fast-growing tree that lives to great age, it is often seen at its best as a feature in an open landscape. When coppiced, to produce its osier withies (strong, flexible twigs used for weaving baskets), the fiery coloured, new young growth is revealed.

Several varieties of the species have been identified and named since early times, and many clones have been developed for ornamental use. *Salix alba* var. *vitellina* is the most commonly grown variety, but perhaps the best known is *S. alba* var. *caerulea*. This is the cricket-bat willow, discovered in Norfolk in the early nineteenth century – a little before the arrival of legendary cricketer W. G. Grace. Once used to make yokes for dairymaids and Sussex trugs (light wooden baskets), this species is now exclusively farmed for its supple timber, which becomes the blade of all modern cricket bats. Although the wood is still grown in the United Kingdom, most bats are now manufactured in Pakistan.

Willows are the natural source of salicin or salicylic acid, first isolated from the bark in 1828. In 1853 the French chemist Charles Frédéric Gerhardt produced from it acetylsalicylic acid (ASA), first trademarked as 'Aspirin', the most commonly produced and frequently prescribed medicine in the world.

Over the centuries the white willow has served the United Kingdom very well indeed. Its timber has been put to such uses as turnery, millwork, coopery and weatherboarding. It has produced fine rafters for the construction of roofs, and its stronger shoots have been used to make tool handles. The bark, which is dark and deeply fissured, is a close second to English oak for tanning. It also makes by far the best charcoal for gunpowder, and as such – in company with the oak – it was a key species in the history of the British Empire. Had he been successful, Guido Fawkes would have had as much cause to praise the white willow in November 1605 as legendary England batsman Len Hutton did at the Oval in August 1938.

Ficus macrophylla (Moraceae)

MORETON BAY FIG

Aboriginal Fishing

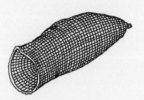

The figs belong to a large genus of mostly woody plants in the mulberry family, Moraceae, of which there are some 850 species worldwide, growing mostly in the tropics. The family also includes the breadfruit and jackfruit. The Moreton Bay fig, a native of Australia, belongs to a group of figs known as banyans. The group includes the Indian banyan tree (see page 86) and the rubber plant, *Ficus elastica*, most familiar as an ornamental houseplant. More often than not, such trees start life as epiphytes (plants that grow on a host plant), and both the Indian banyan and the Moreton Bay fig are referred to as strangler figs. This otherworldly group of species start life in the crown of other trees, the roots growing down from the crown until they reach the soil, and in so doing strangling the host.

The fig is named after Moreton Bay, Queensland, Australia, the tree's native home. Its botanical name, *F. macrophylla*, gives a clue to its appearance – its evergreen leaves (*phylla*) being large (*macro*) and shiny, even leathery, reminiscent of those of the rubber plant. The massive trunks are perhaps the most remarkable of any tree, forming distinctive buttresses from the descending roots, giving the impression that the tree is melting into the soil. The roots thicken as they reach the ground, providing more buttresses to support the immense canopy above. A single tree can take up an entire hectare (2½ acres) of land. It is far too large for any suburban garden, and can damage pavements and the foundations of buildings. In Australia, Aborigines have traditionally used the fibres to make fishing nets, bags and cloth, and the tree is presently the subject of Australian homeopathic research.

Despite being a subtropical rainforest species, it is adaptable to a range of soils and appreciates moisture. The tree grows well in warm, dry, frost-free or Mediterranean climates. It features as a specimen planting in the botanic gardens of Brisbane, Melbourne and Sydney, and there is also a remarkable specimen in the Piazza Marina in Palermo, Sicily. This tree, which is more than 150 years old, is the largest example in Europe, and its girth is impossible to measure because of the tangle of its trunks.

OTHER COMMON NAMES
Australian banyan, strangler fig

ORIGIN
Eastern Australia

CLIMATE AND HABITAT
A very adaptable tree found on
many soil types in subtropical,
warm temperate and dry rainforest;
cultivated in Mediterranean
climates as an ornamental

LONGEVITY
Up to 270 years

SPEED OF GROWTH
60–90 centimetres/
24–35 inches per year

MAXIMUM HEIGHT
60 metres/200 feet

The fruit of the Moreton Bay
fig is small and sweet with
a dry, grainy texture.

OTHER COMMON NAME
Queensland nut

ORIGIN
Queensland, New South
Wales (Australia)

CLIMATE AND HABITAT
Frost-free climate with high rainfall
and humidity; moist, fertile soils

LONGEVITY
50–120 years

SPEED OF GROWTH
30–60 centimetres/
12–24 inches per year

MAXIMUM HEIGHT
20 metres/66 feet

*Similar to the walnut fruit,
the macadamia is green-shelled
and contains a single nut.*

Macadamia tetraphylla (Proteaceae)

MACADAMIA

Adored by the Honey Bee

The Macadamia tree is a member of the *Protea* family, a name borrowed by Carl Linnaeus (the father of botanical taxonomy) from that of the Greek god Proteus, who famously could take many forms. The trees are native to Australia, and the two most commercially important species live in Queensland, from which comes macadamia's local name, 'Queensland nut'. In nineteenth-century Australia, the macadamia was the first crop of native origin to be grown commercially by settlers.

The tree is small but beautiful, with a dense canopy of long, leathery, toothed and wavy-edged leaves in summer and long tassels of pink flowers dripping from its branches in spring. These flowers are much sought after by honey bees, and many nut farmers collaborate with honey producers to maximize the harvest for both. It takes six or seven years from planting for the tree to bear fruit, but the wait is worthwhile, and the nuts are prized for their rich, soft meat. Macadamia fruit is green-shelled, similar to the fruit of the walnut tree. Like walnuts, macadamia shells open to reveal a single extremely hard-shelled edible nut. Next to olives, macadamia kernels are the richest source of cholesterol-lowering monounsaturated fatty acids there is, and are therefore much in demand.

Despite the origin of the tree, and the fact that Australia is one of the largest individual producers, most commercial production by volume takes place elsewhere. The species *Macadamia tetraphylla* now yields an important crop in California and Florida, as well as Mexico, South Africa, Kenya and many other countries. Most commercial nuts come from trees of *M. integrifolia* or hybrids of it, largely because the nuts of this species contain less sugar and are thus less prone to burning during roasting. The more sugary nuts of *M. tetraphylla* are tastier when eaten raw. Ironically, despite the commercial spread of the species worldwide, it is severely threatened in its native land by loss of habitat, caused largely by the destruction of the rainforest to make way for agriculture, and subsequent urban development.

EMPRESS TREE

Grown for Girls

The empress tree in full flower is one of the most magnificent sights. Violet-blue foxglove-like blooms decorate its branches in late spring and early summer, usually before the leaves appear. When seen from a distance the flowers can look like those of the jacaranda tree (see page 192), but its other common name, the foxglove tree, signposts the difference between the two trees, referring to the empress tree's upright flower spikes. The foliage is also remarkable, especially on young trees, whose huge, felted, heart-shaped leaves can be as much as 60 centimetres (2 feet) across.

Paulownia was originally named *Pavlovnia* by the German-born father of Japanese flora, Philipp Franz von Siebold, who is himself commemorated in the botanical names of many Japanese native plants. He named this one after the daughter of the Romanov tsar Paul I of Russia, Anna Pavlovna (born 1795), who married William II, Prince of Orange, and became queen consort of the Netherlands.

So fast does this tree grow that in the state of Texas, where pollution and soil contamination are great problems, it is used as a purification crop. It can absorb ten times more carbon dioxide than any other tree, and expels vast amounts of oxygen. Flourishing in toxic soil, it purifies the land as it matures. A planted seedling will, after only eight years, be the same size as a 40-year-old oak, and in a single year it can grow up to 5 metres (16 feet).

The empress tree has long been cultivated in Japan, where it is known as *kiri* or the princess tree. It is valued there both symbolically and for its fine timber. The wood is closely associated with Japanese female identity. It was once customary to plant a tree when a baby girl was born. The tree would mature with the child, and would be cut down in preparation for her marriage, to be made into wooden articles for her dowry. On her wedding day, her parents would present her with a chest made from her own *kiri* wood, in which to store her kimonos and other fine garments. The tree remains deeply symbolic in Japan to this day. A stylized image of its foliage and flowers is depicted in the seal of the office of the prime minister, and serves as the emblem of the Japanese government.

OTHER COMMON NAMES
Foxglove tree, princess tree, kiri

ORIGIN
China

CLIMATE AND HABITAT
Temperate climate with warmer
summers, but quite adaptable;
the more fertile the soil, the
faster the tree will grow

LONGEVITY
Up to 50 years

SPEED OF GROWTH
80 centimetres–5 metres/
2½–16 feet per year

MAXIMUM HEIGHT
25 metres/82 feet

*The empress tree has violet-
blue foxglove-like flowers.*

The leaves of the western redcedar smell like pineapple or pear drops when crushed.

OTHER COMMON NAMES
Redcedar, western red cedar, western arborvitae, giant arborvitae, Pacific redcedar

ORIGIN
United States Pacific Northwest, Canada

CLIMATE AND HABITAT
A riparian tree, growing in forested swamps and along waterways in its native range; adaptable to drier regions on a wide range of soil types

LONGEVITY
At least 1,400 years

SPEED OF GROWTH
10–30 centimetres/ 4–12 inches per year

MAXIMUM HEIGHT
70 metres/230 feet

Thuja plicata (Cupressaceae)

Western Redcedar

Totem-pole Timber

This giant conifer from the Pacific Northwest of North America is often referred to as western red cedar. It is in fact not a true cedar and is unrelated to the true species of cedar, which belong to the genus *Cedrus* (see page 50). Its wood, however, is similar to that of the true cedars, being light in weight and extremely resistant to rot. The wood of trees that fell more than 100 years ago can still be useably planked. In its native land, the tree can grow to huge proportions, regularly reaching 60 metres (200 feet) in height. The largest specimens grow on Vancouver Island. Elsewhere it is far more likely to be found as a hedging plant, or as a useful alternative to the Leyland cypress, being far better adapted to damp and shade.

The tree is also sometimes referred to as western arborvitae, largely because the other North American species of *Thuja* is known as eastern arborvitae (*T. occidentalis*). The Latin *arbor-vitae*, 'tree of life', refers to its medicinal properties, since it was much valued by the indigenous peoples of North America for treating anything from colds, internal pains and rheumatism to toothache, sore lungs and venereal disease. The western redcedar also has a long history of essential use in northwest coast indigenous culture, and is of great spiritual significance. Some Native Americans from this area refer to themselves as 'the people of the redcedar', owing to their dependence on the tree. The timber is used for housing and is crafted into such objects as utensils, boxes, musical instruments, arrow shafts and such ceremonial objects as masks. It is also the material from which canoes are made (sometimes known by the Greek term *monoxylon* – made from a single piece of wood).

The most notable and noticeable use of the western redcedar is in the making of totem poles. These are found throughout the coastal areas of the Pacific Northwest, in Canada, Washington state, southeastern Alaska and especially in British Columbia (of which province the western redcedar is the arboreal emblem). More than anything, these monumental carvings are a means of communication, commemorating ancestors or symbolizing cultural beliefs and legends, tribal lineages, notable historic events and much more.

Araucaria bidwillii (Araucariaceae)

BUNYA

Harvest Nut

Everything about the bunya pine's appearance suggests that it is a species dinosaurs would have recognized. The rainforest tree looks positively Jurassic, the period in which it first appeared on Earth, and its fruit cones are roughly the size of dinosaur eggs. The seeds that are enclosed in the bunya's giant heavy cone, the size of unshelled Brazil nuts, are extremely nutritious and can be eaten raw or cooked, or ground into flour. The trunk of the tree is coarse with a serrated pattern of horizontal rings. The branches seldom overlap, giving the tree the appearance of a giant bottle-brush. Although it is referred to as bunya pine, it is not a true pine but a conifer, and is most closely related to the monkey puzzle tree (see page 154). It is native to Queensland, Australia, specifically to eastern parts of that state.

Colonial settlers did much to diminish the populations of the bunya pine because of its value as timber. This threatened the traditions of the Aboriginal people for whom the bunya bunya, as it is known to them, was culturally of great significance. Bunya festivals involved some of the largest gatherings of the indigenous peoples, with tribes travelling great distances to celebrate the nut harvest, which lasted for months. It was a moveable feast, the timing of which depended on when the cones fell. It was also an occasion to set aside tribal differences in the interest of the common good. Many such festivals are still held, although their nature has changed and they are now more an excuse for music, food and cultural events.

The species was given its Latin name after John Carne Bidwill, an English botanist who documented much of the flora of Australia and New Zealand in the nineteenth century. He identified the bunya and introduced it to the newly formed Royal Botanic Garden at Kew in 1842.

The Edinburgh-born Australian explorer Thomas Petrie, who was brought as an infant to live in Australia, mixed freely with Aboriginal children and learned to speak their language. At the age of 14 he was invited on a Walkabout to visit a bunya festival in the Bunya Mountains, and his daughter Constance later recorded the event. Her book *Tom Petrie's Reminiscences of Early Queensland*, published in 1904, is regarded as one of the best accounts of early colonial life in Queensland's capital, Brisbane.

*Millions of years ago dinosaurs would
have eaten the bunya's huge cones.*

The copper beech, a popular cultivar of the European beech, has beautiful dark purple leaves.

OTHER COMMON NAMES
Beech, common beech

ORIGIN
Europe, Asia

CLIMATE AND HABITAT
Humid climate; free-draining, calcified or slightly acidic soil

LONGEVITY
At least 500 years

SPEED OF GROWTH
10–60 centimetres/
4–24 inches per year

MAXIMUM HEIGHT
45 metres/148 feet

Fagus sylvatica (Fagaceae)

EUROPEAN BEECH

Leaf Aperitif

The European beech is one of the most magnificent of all forest trees. Those lucky enough to have discovered the world of Winnie the Pooh, Piglet and Christopher Robin will have come across the beech in E. H. Shepard's classic illustrations of the 'Hundred Acre Wood'. The trunk of the beech is tall and strong, its foliage a transparent green in spring, turning to gold and copper in autumn, when its fallen leaves carpet the ground. Its timber is fine, and without doubt one of the best for fire and furniture.

The species is widespread across Europe, but in the United Kingdom it is common only in the southern half of the country, where it is not considered truly native. In continental Europe it is cultivated more widely, and in France, where the tree is common, the young leaves are collected and steeped in gin to make beech-leaf *noyau*, a delicious lime-green aperitif with a nutty flavour.

One of the greatest displays of beech is the Sonian Forest in Belgium, southeast of Brussels, straddling all three regions of the country: Flemish, Walloon and Capital. This magnificent old-growth forest of about 45 square kilometres (17 square miles) is predominantly of beech, with some oak and hornbeam. It is part of what remains of the vast ancient Silva Carbonaria (charcoal forest), and it contains many trees that are now at least 275 years old and still growing.

The European beech has been exported widely to other temperate regions of the world, and there is a huge array of varieties that can be grown as garden ornamentals for foliage colour and habit, far beyond the familiar purple-leaved types. It is very popular in the United States, and was introduced there as an ornamental shade tree that tolerates urban environments far better than does the American native beech, *Fagus grandifolia*. The oldest and largest collection of trees is in the Longwood Mall in Brookline, Massachusetts, a 1-hectare (2.5-acre) arboretum planted in about 1850. European beeches can also be found in other American parks, including Central Park in New York, where there are many fine examples that may have served as a reminder of home to the city's many European immigrants.

Jacaranda mimosifolia (Bignoniaceae)

JACARANDA
The Exam Tree

There is an Amazonian myth concerning the deciduous blue jacaranda. It is said that a bird named Mitu descended from the heavens, bringing a priestess of the moon to the top of such a tree. The priestess climbed down and went to live in a nearby village. There she shared her knowledge with the villagers until Mitu returned to take her back to the heavens and return her to her soulmate, the son of the Sun.

Each day millions of people, especially those who live in Brazil or Argentina, Mexico or South Africa – any country that has little or no frost – will walk along city streets lined with the blue jacaranda. The name 'jacaranda' appears on hotels, restaurants, bars, shops and even radio stations in those cities, thanks to the fame of a tree that offers a fabulous purple display in spring and welcome shade in summer.

There are few trees as breathtakingly beautiful as the blue jacaranda, especially in spring, when it is in full flower. The colour is electric. It is estimated that one million of these trees grow in Pretoria, South Africa, giving the city its other name, Jacaranda City. The trumpet-shaped blossoms turn the city blue from September to November. It is a magical display, but perhaps it is almost too much of a good thing. In Pretoria, the jacaranda is now classified as a weed and the planting of it is restricted, although the tree seems to take little notice of this restriction. Its flowering time coincides with the final exams of university students, and a modern myth has developed: if a jacaranda flower lands on a student's head, that lucky student will pass his or her exams with flying colours.

Pretoria is not alone in celebrating this beautiful tree. In Queensland, Australia, jacaranda festivals are held to mark the coming of spring. There, the tree is often referred to as the 'exam tree', as its flowering coincides with university finals, and the phrase 'purple panic' is used to describe a student's pre-exam state of mind.

OTHER COMMON NAMES
Blue jacaranda, fern tree

ORIGIN
Brazil, Argentina

CLIMATE AND HABITAT
Hot, dry, frost-free climate;
wide range of neutral to
acid free-draining soils

LONGEVITY
Up to 100 years

SPEED OF GROWTH
20–50 centimetres/
8–20 inches per year

MAXIMUM HEIGHT
20 metres/66 feet

*The jacaranda's fabulously coloured
flowers appear in spring and early
summer, and last for two months.*

OTHER COMMON NAMES
Handkerchief tree, ghost tree

ORIGIN
Southwestern and central China

CLIMATE AND HABITAT
Temperate areas; moist,
moderately fertile soil

LONGEVITY
Up to 200 years; the oldest
specimens in cultivation
are now 120 years old

SPEED OF GROWTH
20–50 centimetres/
8–20 inches per year

MAXIMUM HEIGHT
25 metres/82 feet

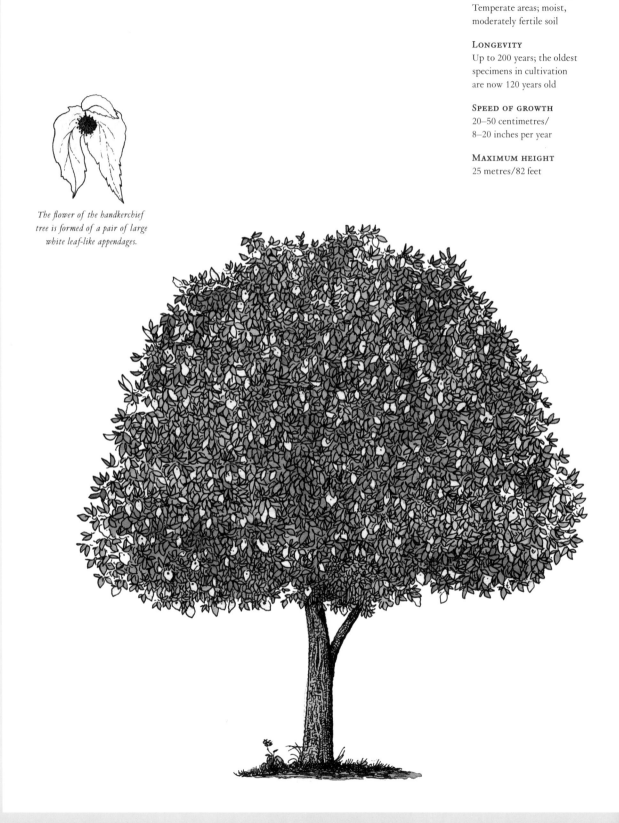

*The flower of the handkerchief
tree is formed of a pair of large
white leaf-like appendages.*

Davidia involucrata (Nyssaceae)

DOVE TREE

Plant-hunter's Prize

The Victorian era in Britain was a golden age for the introduction of new ornamental trees, which were sought avidly at great expense and in some cases personal sacrifice. New trees were important status symbols for the nobility and middle class of the time, and today many special examples survive in the gardens of large estates that were originally privately owned but, in many cases, are now open to the public.

The lust for trees during this era was so great that it triggered the birth of the commercial plant-hunter. One of the most prolific, the young English naturalist Ernest Henry 'Chinese' Wilson, was employed by the famous London nurseryman Sir Harry Veitch, who dispatched him to China specifically to find the dove tree and return with its seed. *Davidia*, however, was discovered by and is named after Father Armand David, a French missionary and keen naturalist who is also commemorated in the botanical name of the Chinese white pine, *Pinus armandii*. David was the first Westerner to describe another rare Chinese endemic species, the giant panda. Although it was David who first described the dove tree in 1869, his specimens were lost in a shipwreck on the Han River. The Irish plant-hunter Augustine Henry found a single tree in 1888 and sent the first dried specimens to the Royal Botanic Gardens at Kew. Under orders to go to China to find Henry's tree, Wilson arrived to find that it had been felled for building. He found further specimens, however, and successfully introduced seed and seedlings to the West.

The dove tree is a very desirable deciduous ornamental tree, best known for its numerous flowers, which appear from late spring. They form dangling ball-like clusters 1–2 centimetres (about ½ inch) across, reddish, each attached to a pair of very large pure-white, leaf-like appendages that perform the function of petals. These modified leaves, known as bracts, flutter in the wind like white doves, and when they fall they look like dropped handkerchiefs, hence the tree's common name.

Citrus × *latifolia* (Rutaceae)

LIME

Mexican Shot

The common name lime is applied to a number of citrus fruits that are, in general, the most acidic. Those of us who cannot grow our own, nor have the luxury of locally grown fruit, are most likely to come across the Persian lime in our culinary experiments or when slicing it for our G&Ts, as this is the most commercially used lime variety. It is favoured for its larger fruit, which is juicier and less acidic than its rival, the Mexican or Key lime (*Citrus* × *aurantiifolia*). Although it is now recognized as a hybrid of Mexican lime and lemon (see page 116), the origin of this important citrus is lost in history, but Asia is known to be the home of both parents, the hybrid attaining its common name from Persia, the hub of ancient trade routes.

A number of cultivated varieties of *C.* × *latifolia* are grown today, and one of the most important is Bearss lime, also known as the seedless and Tahitian lime. This is most probably a selection from a fruit of Tahitian origin made around 1895, in the J. T. Bearss Nursery, Porterville, California. The Bearss lime is a hardier breed with smaller, thin-skinned, deep-green seedless fruit that, in common with many other limes, turn yellow at maturity.

Mexico is a big producer of limes and also the home of tequila, which is distilled in the highlands of Jalisco. The classic way to enjoy the partnership of lime and tequila is to lick a sprinkle of salt from the back of the hand, down a shot of tequila and bite a slice of lime. The salt, it is said, mollifies the spices of Mexican food, and the lime helps to lessen the aftertaste of the alcohol.

In common with many other citrus trees, the Persian lime bears a profusion of fragrant white blossom among its olive-green leaves on densely packed branches. In favourable climates, the tree produces fruit and flowers throughout the year, and seedless fruit is harvested from June to August.

OTHER COMMON NAMES
Tahiti lime, Bearss
lime, Persian lime

ORIGIN
Asia

CLIMATE AND HABITAT
Warm, subtropical or tropical
climates; moist, free-draining soils

LONGEVITY
Up to 100 years

SPEED OF GROWTH
20–60 centimetres/
8–24 inches per year

MAXIMUM HEIGHT
6 metres/20 feet

*The Persian lime is bigger, juicier and
less acidic than the Key lime. It also
has fewer seeds and a thicker skin.*

The Monterey cypress has small,
aromatic, overlapping leaves.

OTHER COMMON NAME
Macrocarpa

ORIGIN
California (United States)

CLIMATE AND HABITAT
Cool, damp maritime climate;
fertile moist soils

LONGEVITY
Up to 250 years

SPEED OF GROWTH
30–60 centimetres/
12–24 inches per year

MAXIMUM HEIGHT
25 metres/82 feet

Cupressus macrocarpa (Cupressaceae)

MONTEREY CYPRESS

California's Lonesome Survivor

On a rocky outcrop along Pebble Beach, California, stands the Lone Cypress, a windswept specimen now supported by guide wires. It is one of the most photographed trees in North America. Its species, the Monterey cypress, is one of those trees that, like the cedar of Lebanon (see page 50), has an iconic, instantly recognizable shape. The slightly inclined, broad horizontal branching is unmistakable; older specimens are often flat-topped, forcibly carved to this shape by years of abuse from the wind (ironically, it is still used elsewhere as a windbreak tree), and also showing gnarled and twisted trunks that make the trees appear truly ancient. In truth, despite this appearance – and although legends abound about specimens living up to 2,000 years – there is little real evidence that the Monterey cypress lives much beyond a couple of hundred years. The tree does grow astonishingly fast, quickly reaching its large mature size; indeed, it is one of the parents of the ubiquitous hybrid the Leyland cypress, *Cupressus × leylandii*.

This tree is truly native only to a very restricted area on the coast of central California, where two populations still exist to the north and south of Carmel Bay, culminating at Cypress Point to the north. Despite its diminished native range, the Monterey cypress has been widely planted and naturalized elsewhere in the world. It has been cultivated very successfully in parts of Australia and New Zealand, where it has naturalized in some areas. In New Zealand, during the early years of the last century, it was the tree of choice to protect coastal farms from the sea – and it went on to be widely used in farm forestry. Today, known simply as 'macrocarpa', it is still grown for forestry in New Zealand, but its use is becoming more restricted owing to a fungal canker disease that seems more prevalent on the warmer, drier inland plantations. The reason for this is still being researched, but it is fairly clear that this cypress is happiest in cooler zones; much like humans, when trees are happy and healthy they can fend off disease.

Aesculus hippocastanum (Sapindaceae)

HORSE CHESTNUT

The Children's Tree

The conker tree – as it is known in Britain thanks to the children's game played with the seeds or 'conkers' – is a magnificent ornamental shade tree with spreading branches and a rounded crown. The spring flowers, which explode from swollen brown buds, are erect candle-like spikes made up of a pyramid of many individual flowers, each white with red blotches to the upper petals, borne at the ends of the branches. A mature tree in full flower in May is a treat to behold, and that is one of the primary reasons why this native of a relatively small area of the Pindus and Balkan mountains in southeastern Europe has now naturalized across much of Europe, being widely planted in parks and estates, and on streets.

The beer gardens of Germany, particularly Bavaria, were widely planted with horse chestnut trees, which were used to shade the lagering cellars before the advent of modern refrigeration, and now create the shade under which a few steins of amber ale can be enjoyed. Horse chestnut trees have also been widely planted in cities and parks in the United States and Canada.

One very famous European example grew outside the seventeenth-century canal house in Amsterdam where, in an annex of secret rooms, the German-born diarist Anne Frank and her family famously hid from the Nazis during World War II. The tree was of great importance to her, and she wrote of it several times in her *Diary of a Young Girl*. 'The Anne Frank tree', as it became known, survived the war. After suffering from disease for a number of years, it was condemned in November 2007 and scheduled to be cut down. Emotions ran high, however, and the tree was saved by a court injunction. A charitable foundation was set up to save the tree, but it was dealt a cruel blow when an August storm blew it over. There was some early hope that a branch from near the base of the tree would regenerate, but it was finally declared dead in 2010. Happily, seed saved from the tree was sent to the United States, where eleven saplings were raised and distributed for planting at Holocaust remembrance centres, including parks, museums and schools.

OTHER COMMON NAME
Conker tree

ORIGIN
Southeastern Europe

CLIMATE AND HABITAT
Cold to warm temperate
climate; deep, fertile soil

LONGEVITY
Up to 300 years

SPEED OF GROWTH
50–80 centimetres/
20–32 inches per year

MAXIMUM HEIGHT
40 metres/130 feet

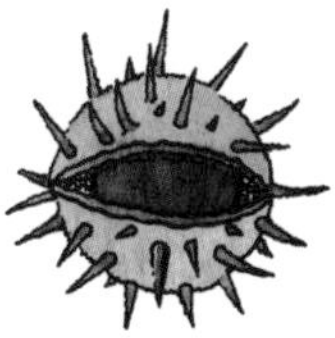

*Conkers got their name from the
word 'conch', as the game was
originally played using snail shells.*

OTHER COMMON NAME
Shui-shan (water fir)

ORIGIN
Hubei Province (China)

CLIMATE AND HABITAT
Temperate moist areas (riverbanks,
floodplains or areas of high rainfall)

LONGEVITY
Up to 600 years

SPEED OF GROWTH
30 centimetres–1 metre/
1–3 feet per year;
commonly achieves 30
metres/98 feet in 60 years

MAXIMUM HEIGHT
60 metres/200 feet

*The dawn redwood is the only
living species of its genus. It is
deciduous, not evergreen, meaning
that it sheds its leaves in autumn.*

Metasequoia glyptostroboides (Cupressaceae)

DAWN REDWOOD

Living Fossil

Although its remarkable pyramidal winter silhouette and wide, gnarled buttress are now familiar, the dawn redwood was known only as a fossil until World War II. The genus was first described in 1941 as a fossil of the Mesozoic era, discovered by a Japanese palaeobotanist from Kyoto University, Dr Shigeru Miki, who identified it while studying fossil samples. He realized that he was looking at a new genus, which he named *Metasequoia* (Sequoia-like). That same year a Chinese forester named T. Kan chanced on an enormous living specimen in his native country and, although unaware of Miki's new genus, recognized a unique tree. It formed part of a local shrine, where villagers called it *shui-shan*, water fir. News of the discovery reached the United States, and in 1948 a seed-collecting team was sent to China by the Arnold Arboretum of Harvard University, from where it was distributed to botanic gardens around the world.

Not only was *Metasequoia* a living relic, previously thought to have become extinct with the dinosaurs, but it also happened to be highly ornamental, and it revived an interest in trees that had been suppressed during the war. In its early revival, its rarity coupled with its beauty created an eagerness among plantsmen to increase its numbers. As *Metasequoia* became better known it revealed just how beautiful it was as a specimen tree. The ferny foliage is a bright emerald-green in spring, turning russet-brown with tints of coppery pink or even fiery red in autumn. Its trunk is particularly remarkable, being very wide at the base, the rough bark ruddy and deeply furrowed and contorted, often with green mosses growing within the crevices. These gnarled buttresses are architectural and highly ornamental.

The tree is very easy to cultivate, and although its eventual size might limit planting in smaller gardens, it can make a superb subject for bonsai. It is not well adapted to drought, but it is very tolerant of maritime exposure and pollution, and therefore makes a very good street tree. The city of Pizhou in China has the longest dawn redwood avenue in the world. Originally 60 kilometres (just over 37 miles) long, it contained more than a million trees when it was first planted in 1975. At 5 million trees, the city now boasts the largest number of dawn redwoods planted anywhere.

Betula pendula (Betulaceae)

Silver Birch

Silver Lady of the North

The elegant silver birch – often known simply as 'lady of the woods' – is familiar to many for its smooth white trunk, making it unmistakable, especially in winter. Native across the whole of Europe and into Asia, it is nowhere more at home than in the Caledonian forests of the Scottish highlands, where it thrives on the poor, acidic soil. These are true wildwood forests that have changed little since they were formed after the last ice age, and, along with the Scots pine (see page 128), the silver birch is still dominant there.

Birch rarely lives to a great age, but where it is happy it grows and spreads quickly. Its smooth, thin, dark purplish-brown twigs have what look like small warts. The buds open in April, revealing long-stalked serrated-edged leaves with pointed tips. These leaves mature from pale to dark green, turning yellow in early autumn. When felled, the fine yellowish-white timber is hard enough for general carpentry, if slightly perishable. Walk through a birch wood and you will see fallen, hollowed-out trunks where the wood has rotted away, leaving the bark intact and making great hiding places for many woodland creatures. This demonstrates the porous nature of the wood, which in times past was tapped for its sap to make wine – a drink that is occasionally still produced by foragers. The bark, which contains a very strong resin that was once used for tannin and glue, endures, and – being waterproof – was traditionally used as roof shingles.

The birch is not only beautiful, but has also been highly useful to humans throughout our existence. In the harsh winter climate of northern Europe its tough hardwood was needed in the past for warmth, shelter, footwear, healing and drink; its long, thin, cord-like branches are still the main component of besom brooms; and in Finland, where it has been the national tree since 1988, those same branches, usually in leaf, are used in the sauna for gentle self-flagellation, said to act as a muscle relaxant and to soothe the irritation caused by mosquito bites. Birch is commercially most important in Scandinavia, where its vast forests are the main source of pulp for paper-making.

OTHER COMMON NAME
Lady of the woods

ORIGIN
Europe, Asia

CLIMATE AND HABITAT
Very adaptable; happy in all
but the most severe drought
or flooded areas, on acid soil,
particularly heathland

LONGEVITY
50–100 years

SPEED OF GROWTH
10–80 centimetres/
4–32 inches per year

MAXIMUM HEIGHT
30 metres/98 feet

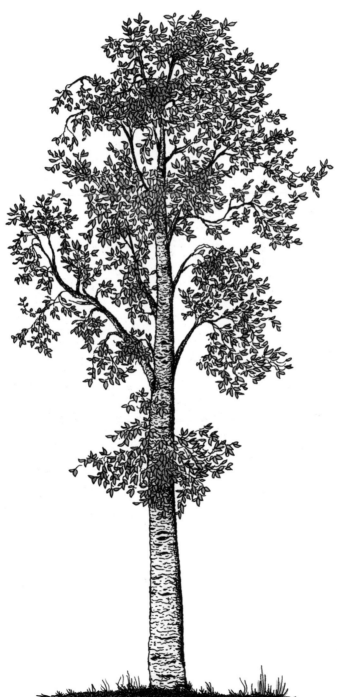

*In Finland leaves of the silver
birch tree are used in the
preparation of a popular tea.*

OTHER COMMON NAME
Silver wattle

ORIGIN
New South Wales, Victoria,
Tasmania (Australia)

CLIMATE AND HABITAT
Very adaptable; happy in
all but the wettest areas

LONGEVITY
No more than 30 years

SPEED OF GROWTH
60 centimetres–2 metres/
2–6½ feet per years

MAXIMUM HEIGHT
30 metres/98 feet

*The spectacular mimosa flowers
in very early spring with panicles
of bright yellow pompoms.*

Acacia dealbata (Fabaceae)

MIMOSA

A Flower to Empower

Much as the red rose has become symbolic of Valentine's Day, the mimosa tree, and in particular its blossom, has in recent times become a symbol of International Women's Day, which is celebrated particularly enthusiastically in Italy. La Festa della Donna, as it is known there, is celebrated on 8 March each year, when flowering branches of mimosa are given and received by women as a symbol of solidarity.

It is not surprising that the mimosa was chosen for this purpose. With its abundant panicles of yellow pompoms, it spreads the most joyous scents and sights of early spring, sending a strong hint to winter that its days are numbered. Understandably, bees and insects love the flowers too and, especially for bees, they are a valuable source of late winter fuel. When not overwhelmed by brilliant blossom, the pinnate leaves can be seen, large and feathery, usually blue-green but occasionally a silvery green.

The tree is fairly common across southern Europe, where it has naturalized in some places. It was probably introduced for ornamental purposes, and the flowers and foliage are valued highly in the cut-flower trade. In southern France, oil produced from the flowers is used as a fixative in high-grade perfume production, and the sensual fragrance from the flowers is also manufactured into an absolute (concentrated aromatic similar to an essential oil) known as Cassie, which is also produced from another species, *Acacia farnesiana*, a thorny relative of the mimosa.

Wherever the temperature permits – since -10ºC (14ºF) is about as cold as it will stand – the mimosa can be grown as a very fine moderately sized ornamental tree. The flowers appear in such abundance that they blanket the exquisite, soft, finely cut fern-like foliage and then, as they fall, cover the ground beneath in a carpet of gold. This is a very fast-growing tree, but it is not especially long-lived. In its native Australia it is regularly subjected to bush fires, which it survives not only by regenerating from the root, but also because the flames cause the hard seed coat of the long-dormant seeds in the soil to crack, enabling them to germinate as soon as moisture permits. This pioneer species is therefore one of the first trees to recolonize scorched earth.

Pyrus calleryana (Rosaceae)

CALLERY PEAR

The Survivor

There is nothing remarkable about the Callery pear in ornamental terms, but it is one of the most frequently planted street trees. It was chosen as a virtually indestructible species for use in urban planting schemes, being tolerant of pollution and heavy, compacted soils. Its resilience was proven by the remarkable discovery, after the terror attack of 11 September 2001 on the World Trade Center in New York, of a charred Callery pear stump at Ground Zero. It was removed, nurtured back to health and replanted there ten years later, and now stands once again, resplendent, having earned the moniker 'Survivor Tree'.

The Callery pear was identified in 1872 by the French missionary Joseph-Marie Callery – for whom it is named – on a visit to China. It was introduced to the United Kingdom by biologist and plant collector Ernest Henry Wilson in 1908, one of 2,000 species he brought to the West in the late nineteenth and early twentieth centuries. The tree inspired little interest until it was brought to the United States by botanical explorer Frank Meyer in 1918. It was one of the last plants Meyer introduced before his untimely and mysterious death in June that year, when he fell from a Japanese riverboat into the Yangtze River.

Pears are among the most important fruit trees in the northern hemisphere, and the genus, *Pyrus*, is a large one. The wood of all species is close-grained and durable, prized especially for woodwind instruments. Varieties of the Callery pear were selected for prolific flowering and autumn leaf colour. 'Bradford' is one such clone, which became almost ubiquitous across much of urban North America.

Pyrus calleryana seemed perfectly suited to its habitat, but as it ages it becomes highly susceptible to storm damage, and has flowers which en-masse have an unpleasant odour. There are places in North America where this pear has escaped urban captivity to start a new life in the wild, to the detriment of the native flora. Callery would surely have preferred to be remembered for his monumental and important *Encyclopedia of the Chinese Language*, the first of its kind, published in 1842.

OTHER COMMON NAME
Bradford pear

ORIGIN
China, Vietnam, Korea, Japan

CLIMATE AND HABITAT
Very adaptable; happy in all but the
driest or most waterlogged areas

LONGEVITY
Up to 100 years

SPEED OF GROWTH
20–50 centimetres/
8–20 inches per year

MAXIMUM HEIGHT
15 metres/49 feet

*The Callery pear produces clusters of
beautiful white flowers followed by
brown fruits, which have a bitter taste.*

OTHER COMMON NAMES
None

ORIGIN
Peru, Ecuador

CLIMATE AND HABITAT
Wet montane cloud forests at
1,800–2,400 metres/6,000–7,800
feet above sea level

LONGEVITY
Unrecorded

SPEED OF GROWTH
5–10 centimetres/
2–4 inches per year

MAXIMUM HEIGHT
25 metres/82 feet

*The leaf of the elusive Esser's tree of
the Inca is large, shiny and leathery.*

Incadendron esseri (Euphorbiaceae)

ESSER'S TREE OF THE INCA

New Discovery

Esser's tree of the Inca is a fitting way to bring our story to a close. Hiding in the cloud-covered forest of the High Andes of Peru, and not officially named or described until 2017, it demonstrates just how much more there may be for us to discover about the uniquely fascinating history of the tree.

Incadendron esseri was found by researchers from the Smithsonian Institution and Wake Forest University, studying the unique ecology of this part of Peru. It was found to be common within a fairly narrow altitudinal range in the cloud forest zone of one of the world's most biodiverse regions, an area that is remote but not inaccessible. It was first seen growing along the famous Trocha Union trail, which was built and used by the sun-worshipping Inca civilization and leads from Tres Cruces, a famous sunrise mirador at 3,600 metres (11,800 feet), down to the bread basket of the lower Andes and the mighty Amazon far below.

The tree itself is full of arboreal character, with curling and sometimes pendulous branches that wind their way out of the twisting, scaly trunk, and large, shiny, leathery leaves similar to those of the laurel. Esser's tree is a member of the spurge family, Euphorbiaceae, a large family that includes an incredibly wide variety of species, from the rubber tree (see page 64) and cassava to the castor oil plant, and from the candlenut tree (see page 110) to the poinsettia. Although diverse in appearance, several members of this family share a common characteristic: when damaged they exude a sticky latex. Had Esser's tree been discovered in an earlier age, it might well have competed with the rubber plant.

Perhaps we should end with the words of Kenneth Wurdack, a botanist at the Smithsonian National Museum of Natural History: 'This tree perplexed researchers for several years before being named as new. It just goes to show that so much biodiversity is unknown and that new species are awaiting discovery everywhere – in remote ecological plots, as well as in our own backyards.'

List of World Botanical Gardens and Arboreta

The historical origin of the botanic garden dates back to the European medieval medicinal gardens, known as physic gardens, which were started in the time of Charlemagne in the eighth century. The Chelsea Physic Garden in London, originally the Garden of the Society of Apothecaries, is perhaps the most famous of these, and was founded in 1673 to study the medicinal qualities of plants, becoming one of the most important centres of botany and plant exchange in the world.

Botanical gardens are dedicated to the collection, growing and display of a huge array of plants, labelled with their botanical names, and have seen a revival in recent years due to the emergence of the conservation movement, with growing fears of climate change. They are now seen as hugely important because of their existing collections, and the knowledge they possess in the conservation and propagation of plant species. Some are also centres of scientific research, such as the Royal Botanic Gardens at Kew in London.

There are over 1,700 botanic gardens in 148 countries around the world. Such is their importance that many more are currently under construction.

The following list is just a sample of what the authors believe to be some of the most significant collections, and in particular include those with arboretums (tree collections). A useful resource to find others that may be closer to visit can be found at Botanic Gardens Conservation International (www.bgci.org).

ARGENTINA

BUENOS AIRES BOTANICAL GARDEN
Buenos Aires

GRIGADALE ARBORETUM
Buenos Aires

AUSTRALIA

GALLOP BOTANIC RESERVE
Cooktown

ROYAL BOTANIC GARDENS SYDNEY
Sydney

ROYAL BOTANIC GARDENS VICTORIA
Melbourne

THE NATIONAL ARBORETUM
Canberra

THE TASMANIAN ARBORETUM
Devonport, Tasmania

AUSTRIA

UNIVERSITÄT WIEN BOTANISCHER GARTEN
Vienna

BELGIUM

ARBORETUM TERVUREN
Tervuren

ARBORETUM WESPELAAR
Haacht

ARBORETUM KALMTHOUT
Antwerp

BRAZIL

JARDIM BOTÂNICO
Rio de Janeiro

CANADA

JARDIN BOTANIQUE DE MONTRÉAL
Montreal

ROYAL BOTANICAL GARDENS ONTARIO
Burlington

UNIVERSITY OF BRITISH COLUMBIA (UBC) BOTANICAL GARDEN
Vancouver

CHILE

JARDÍN BOTÁNICO DE LA UNIVERSIDAD AUSTRAL DE CHILE
Valdivia

CHINA

BEIJING BOTANICAL GARDEN
Beijing

KUNMING BOTANICAL GARDEN
Kunming

SHANGHAI BOTANICAL GARDEN
Shanghai

CZECH REPUBLIC

BOTANICAL GARDEN AND ARBORETUM OF MENDEL UNIVERSITY
Brno

FRANCE

ARBORETUM DE BALAINE
Auvergne

L'ARBORETUM NATIONAL DES BARRES
Montargis

ARBORETUM DES GRANDES BRUYÈRES
Orléans

GERMANY

BOTANISCHER GARTEN MÜNCHEN
Munich

ELLERHOOP-THIENSEN ARBORETUM
Ellerhoop

FORSTBOTANISCHER GARTEN UND
ARBORETUM
Göttingen

SPÄTH-ARBORETUM
Berlin

HONG KONG

SHING MUN ARBORETUM
Hong Kong

ISRAEL

JERUSALEM BOTANICAL GARDEN
Jerusalem

ITALY

GIARDINI BOTANICI HANBURY
La Mortola, Liguria

ORTO BOTANICO DI PADOVA
Padua

MEXICO

JARDÍN BOTÁNICO DE LA
UNIVERSIDAD AUTÓNOMA DE
PUEBLA
Puebla

JARDÍN BOTÁNICO DEL INSTITUTO
DE BIOLOGÍA (UNAM)
Mexico City

JARDÍN BOTÁNICO FRANCISCO
JAVIER CLAVIJERO
Xalapa

NEPAL

NATIONAL BOTANICAL GARDEN
KATHMANDU
Kathmandu

THE NETHERLANDS

ARBORETUM TROMPENBURG
Rotterdam

BELMONTE ARBORETUM
Wageningen

NEW ZEALAND

EASTWOODHILL ARBORETUM
Gisborne

GONDWANA ARBORETUM,
AUCKLAND BOTANIC GARDENS
Auckland

HACKFALLS ARBORETUM
Gisborne

OMAN

OMAN BOTANIC GARDEN
Muscat

POLAND

KÓRNIK ARBORETUM
Kórnik

WOJSŁAWICE ARBORETUM
Wojsławice

PORTUGAL

BUÇACO FOREST – MATA
NACIONAL DO BUÇACO
Luso

SINGAPORE

SINGAPORE BOTANIC GARDENS
Singapore

SOUTH AFRICA

KIRSTENBOSCH NATIONAL
BOTANICAL GARDENS
Cape Town

SPAIN

ARBORETUM DE MASJOAN
Girona

ARBORETUM LA ALFAGUARA
Granada

SWEDEN

UNIVERSITY OF UPPSALA
BOTANICAL GARDEN
Uppsala

UNITED ARAB EMIRATES

SHARJAH BOTANICAL GARDENS
Sharjah

UNITED KINGDOM

ABNEY PARK ARBORETUM
London

BEDGEBURY NATIONAL PINETUM
AND FOREST
Bedgebury, Kent

CHELSEA PHYSIC GARDEN
London

ROYAL BOTANIC GARDENS
EDINBURGH
Edinburgh

ROYAL BOTANIC GARDENS AT KEW
London

SIR HAROLD HILLIER GARDENS
AND ARBORETUM
Romsey, Hampshire

WESTONBIRT ARBORETUM
Westonbirt, Gloucestershire

UNITED STATES OF AMERICA

THE ARNOLD ARBORETUM
AT HARVARD UNIVERSITY
Boston, MA

HAWAII TROPICAL BOTANICAL
GARDEN
Hawaii, HI

HOYT ARBORETUM
Portland, OR

JC RAULSTON ARBORETUM
AT NORTH CAROLINA STATE
UNIVERSITY
Raleigh, NC

LONGWOOD GARDENS
Philadelphia, PA

LYON ARBORETUM
Hawaii, HI

MISSOURI BOTANICAL GARDEN
St. Louis, IL

MORRIS ARBORETUM OF THE
UNIVERSITY OF PENNSYLVANIA
Philadelphia, PA

NEW YORK BOTANICAL GARDEN
New York, NY

UC DAVIS ARBORETUM AND
PUBLIC GARDEN
Davis, CA

THE UNITED STATES NATIONAL
ARBORETUM
Washington D.C.

UNIVERSITY OF CALIFORNIA
BOTANICAL GARDEN
Berkeley, CA

Index

Note: common names are printed in plain type; species names are printed in italics; family names are printed in capitals.

The Authors

Kevin Hobbs is a professional grower and plantsman with over 30 years' experience in the horticultural industry. He was Head of Research and Development at world-renowned Hillier Nurseries in Hampshire, UK, and currently works for Whetman Plants International on new plant development. He has advised on behalf of the Queen at Frogmore House and grew the plants for the 2012 Olympic Park design by Piet Oudolf. He is the co-author of *The Hillier's Gardener's Guides: Herbaceous Perennials*.

David West has been growing trees for 35 years, having trained at Hillier Nurseries, and is consequently a self-confessed treehugger. He runs his own nursery business, specializing in the commercial production of rare trees and other unusual plants, through a mail order website, PlantsToPlant.com, which aims for 'conservation through production' and to make unusual plants more accessible.

Kevin's thanks
With thanks to so many in archaeobotany, botany and pomology for their exhaustive dedication. To the Royal Botanic Garden at Kew, IUCN Red List, the Eden Project, James Armitage of the Royal Horticultural Society, and Roy Lancaster CBE, VMH, my life-long mentor. To Jenni for typing up my hand-written words and to all my family and friends for their support and encouragement. And, finally, to everyone involved in publishing this book, especially my first and enduring plant friend, Dave West, for joining me in this work, and Dr. Alexandra Wagstaffe from the Eden Project who joins us in our celebration of trees.

David's thanks
I would like to thank my editors, designer and illustrator, my long-suffering wife Fiona for her encouragement and patient reading and checking of my text, and my family more generally for putting up with my absence from family life so much during the process. A big thank you to Kevin Hobbs for offering me the opportunity to write about my passion, and to Mark Fletcher for making this book possible. Special thanks also to Alexandra for the wonderful foreword, which summarises perfectly our joint story-telling journey. I would also like to thank the Hillier Nurseries, where I cut my teeth and first developed an interest in trees, and the Sir Harold Hillier Gardens for being the best tree study room in which to indulge the passion. Particular thanks go to Roy Lancaster for his ongoing inspiration and encouragement.